Building a Modern Japan

Science, Technology, and Medicine in the Meiji Era and Beyond

Edited by Morris Low

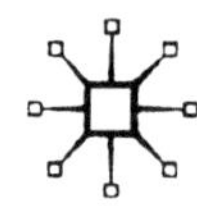

BUILDING A MODERN JAPAN

 A version of chapter 2 was published in the *Journal of the Royal Asiatic Society* (Vol. 15, no. 1 [2005]). We thank the society for permission to reprint it here.

First published in 2005 by
PALGRAVE MACMILLAN™
175 Fifth Avenue, New York, N.Y. 10010 and
Houndmills, Basingstoke, Hampshire, England RG21 6XS
Companies and representatives throughout the world.

PALGRAVE MACMILLAN is the global academic imprint of the Palgrave Macmillan division of St. Martin's Press, LLC and of Palgrave Macmillan Ltd. Macmillan® is a registered trademark in the United States, United Kingdom and other countries. Palgrave is a registered trademark in the European Union and other countries.

ISBN 1–4039–6832–2 hardback

Library of Congress Cataloging-in-Publication Data

Building a modern Japan : science, technology, and medicine in the Meiji era and beyond / edited by Morris Low.
p. cm.
Includes bibliographical references and index.
ISBN 1–4039–6832–2
1. Japan—History—1869– 2. Medicine—Japan—History. 3. Science—Japan—History. I. Low, Morris.

DS881.9.B85 2005
610′.952—dc22 2004062063

A catalogue record for this book is available from the British Library.

Design by Newgen Imaging Systems (P) Ltd., Chennai, India.

First edition: May 2005

10 9 8 7 6 5 4 3 2 1

Printed in the United States of America.

Contents

List of Illustrations vii

Notes on Contributors ix

Preface xiii

Introduction 1
Morris Low

Part 1
Science, Medicine, and a Healthy Nation 11

1. The Rise of Western "Scientific Medicine" in Japan: Bacteriology and Beriberi 13
Christian Oberländer

2. Male Anxieties: Nerve Force, Nation, and the Power of Sexual Knowledge 37
Sabine Frühstück

3. The Female Body and Eugenic Thought in Meiji Japan 61
Sumiko Otsubo

4. Racializing Bodies through Science in Meiji Japan: The Rise of Race-Based Research in Gynecology 83
Yuki Terazawa

5. Doctors, Disease, and Development: Engineering Colonial Public Health in Southern Manchuria, 1905–1926 103
Robert John Perrins

Part 2
Technology, Industry, and Nation 133

6. The Mechanization of Japan's Silk Industry and the Quest for Progress and Civilization, 1870–1880 135
David G. Wittner

7. A Miracle of Industry: The Struggle to Produce Sheet Glass in Modernizing Japan 161
Martha Chaiklin

8. Modernity and Carpenters: *Daiku* Technique and Meiji Technocracy 183
Gregory Clancey

9. The Impact of the Great Depression: The Japan Spinners Association, 1927–1936 207
W. Miles Fletcher III

Index 233

List of Illustrations

Figures

2.1 A pamphlet advertising a 20-volume, abridged Japanese version of Havelock Ellis's *Studies in the Psychology of Sex* 38
2.2 An advertisement for techniques to lengthen the body promising that "Even Short Men Will Become Tall" 46
2.3 The home medical handbook *Katei ryōhō hyakka jiten* (1952) 53
5.1 Manchuria and the South Manchuria Railway 105
5.2 Planned layout of Dairen, ca. late 1910s 111
5.3 The New South Manchuria Railway Hospital, ca. 1926 120
5.4 Floor plans of the New South Manchuria Railway Hospital in Dairen, ca. 1926 121
6.1 Detail of samples of a filature's exterior 147
6.2 *Chambon* method of croisure 148
6.3 A *tavelle* 149
7.1 Production by the cylinder method 172
7.2 Iwasaki Toshiya 173
7.3 Amagasaki Factory 174

Tables

4.1 Average age of menarche for Japanese, Ainu, Ryūkyūan, and Chinese women 92
4.2 Average age of menarche for Japanese women of different categories at the Kumamoto Prefectural Hospital 94

Notes on Contributors

Morris Low, the editor, is Assistant Professor at Johns Hopkins University where he is the Bo Jung and Soon Young Kim Professor of East Asian Sciences and Technology. He has previously taught and conducted research at the University of Queensland, the Australian National University, and Monash University. He is coauthor of *Science, Technology and Society in Contemporary Japan* (CUP, 1999) and coeditor of *Asian Masculinities: The Meaning and Practice of Manhood in China and Japan* (Routledge Curzon, 2003). He has edited special issues of several journals including *Osiris* (University of Chicago Press), *History and Anthropology* (Harwood), *Asian Studies Review* (Blackwell), and *History and Technology* (Taylor and Francis). His three-volume anthology *Science, Technology and R&D in Japan* (Routledge, 2001) has attracted considerable attention.

Martha Chaiklin is Curator of Asian History at the Milwaukee Public Museum. Her Ph.D. was awarded from the University of Leiden in the Netherlands. She is the author of *Cultural Commerce and Dutch Commercial Culture—The Influence of European Material Culture on Japan, 1700–1850* (Leiden: CNWS, 2003) and several articles and translations.

Gregory Clancey is Assistant Professor of History at the National University of Singapore. He has been a Fulbright Graduate Fellow at the University of Tokyo, and a Lars Hierta Fellow at the Royal Institute of Technology in Stockholm. Clancey has coedited *Major Problems in the History of American Technology* (Boston: Houghton-Mifflin, 1998) and *Historical Perspectives on East Asian Science, Technology, and Medicine* (Singapore: Singapore University Press & World Scientific, 2002). His most recent book, *Foreign Knowledge: The Cultural Economy of Japanese Earthquakes*, is forthcoming from the University of California Press.

W. Miles Fletcher III received his M.A. in East Asian Studies and his Ph.D. in History in 1975 from Yale University, where he specialized in modern Japanese history. He is now Professor of History at the University of North Carolina at Chapel Hill where he has served as chair of the Curriculum in Asian Studies. His first book, *The Search for a New Order: Intellectuals and Fascism in Prewar Japan* (University of North Carolina Press, 1982), focused on intellectual and political history, while his second monograph, *The Japanese Business Community and National Trade Policy, 1920–1942* (University of North Carolina Press, 1989) dealt with business history. Since that time, his research has centred on the history of Japanese industrialization with a focus on the Japanese textile industry, about which he has written several articles. His current project examines the recovery of that sector after the Pacific War.

Sabine Frühstück is Associate Professor of Modern Japanese Cultural Studies at the University of California, Santa Barbara. Her academic interests include modern Japanese cultural studies, history and anthropology, the theory and history of sexuality and gender, knowledge systems, post/colonialism, and military–societal relations. She is the author of *Colonizing Sex: Sexology and Social Control in Modern Japan* (2003), and the coeditor of *The Culture of Japan as Seen through Its Leisure* (1998), and *Neue Geschichte der Sexualität: Beispiele aus Ostasien und Zentraleuropa* (1999). Frühstück has also published in *Journal of Asian Studies, Journal of Japanese Studies, American Ethnologist and Jinbun Gakuhō*, among others. She is currently completing a monograph on the armed forces, *Japan Avant-garde: The Army of the Future.*

Christian Oberländer is Professor in the Department for Japanese Studies at the University Halle-Wittenberg and Visiting Scientist at The University of Tokyo. He previously taught at the University of Bonn. His publications include *Between Tradition and Modernity: The Movement for the Preservation of Kanpo Medicine in Japan* (Stuttgart: Steiner, 1995), and *Technology and Innovation in Japan* coedited with Martin Hemmert (London: Routledge, 1998).

Sumiko Otsubo completed her Ph.D. at Ohio State University and has been a postdoctoral fellow at the Reischauer Institute, Harvard University, and taught at Creighton University in Omaha, Nebraska. Currently she is Assistant Professor in the Department of History, Metropolitan State University in St. Paul, Minnesota. Publications include "Eugenics in Japan: Some Ironies of Modernity, 1883–1945," (with James R. Bartholomew), *Science in Context* (Vol. 11, nos. 3–4 [1998]); and "Feminist Maternal Eugenics in Wartime Japan," *U.S.-Japan Women's Journal* (English Supplement, no. 17 [1999]).

Robert Perrins is Associate Professor in the Department of History and Classics, at Acadia University in Nova Scotia, Canada, where he is also the Director of the university's Northeast Asia Research Centre. He completed his Ph.D. in modern Chinese at York University, Toronto. His doctoral thesis was entitled: " 'Great Connections': The Creation of a City, Dalian, 1905–1931: China and Japan on the Liaodong Peninsula." He has been the editor of the China Facts and Figures Annual Handbook (published by Academic Press International), since 1999. He is currently completing a project on the history of disease and Japanese colonial medicine in Manchuria under the sponsorship of the Hannah Institute for the History of Medicine.

Yuki Terazawa received her Ph.D. from the University of California, Los Angeles, where she completed her dissertation entitled, "Gender, Knowledge, and Power: Reproductive Medicine in Japan, 1690–1930." She currently teaches at Hofstra University in Hempstead, New York.

David Wittner completed his Ph.D. at Ohio State University. He is Associate Professor in the Department of History at Utica College, New York. His recent publications include "Chilling Before the Blast: A Comparative Case Study in Technology Transfer in the American and Japanese Iron Industries," *Kinzoku kōzan kenkyū* (no. 77 [May 2000]) and *Commodore Matthew C. Perry and the Perry Expedition to Japan*, The Library of American Lives and Times (New York: Rosen Publishing Group, 2004).

Preface

A special report in *The Economist* of April 10, 2004 was ironically entitled "(Still) Made in Japan." The article related how Japan's manufacturing continued to move overseas, especially to China. The Ministry of Economy, Trade and Industry had recently encouraged the electronics giant NEC to sell its plasma-display business to a local firm, Pioneer, rather than to a foreign company and risk the technology being transferred abroad.[1] While the booming trade within Asia was something to be celebrated, there was real concern that Japan needed to hold on to some of its know-how.

There was, however, some cause for celebration. In February 2004, Japan's trade surplus with Asia exceeded that with the United States for the first time. For the first time in almost a decade, Japan registered a trade surplus with China. Sales of electronic components, semiconductors, and general machinery are said to have been key factors in the improved figures.[2] This has led some commentators to write of "Asia's Eclipse of the West" and the emergence of an Asian-dominated world order centered on China.[3]

Back in the late nineteenth century, the reverse occurred. Japan left what had been a China-centered East Asian system to enter a global system of knowledge dominated by the West. This book throws light on that earlier period of globalization when the Japanese dealt with the new ideas and concepts, and the technologies that were introduced from the West. The first part deals with the role of science and medicine in creating a healthy nation. The second part is devoted to examining the role of technology, and business–state relations in building a modern Japan. It will be seen that it was not a simple process of direct translation of Western know-how. The introduction of Western forms of knowledge certainly helped Japan enter into trade, participate in the international scientific community, and build Western-style buildings, but the way these methods, theories, and systems of knowledge were taken up was arguably "Japanese." As Tessa Morris-Suzuki has argued, local content and cultural heritage helped create a difference.[4]

The Meiji era (1868–1912) has long fascinated students of Japanese history. This volume revisits that period and the decades immediately after. The book effectively brings together state-of-the-art scholarship on the impact of science and technology in Japan's modernization, with studies concerning medicine and the health of the Japanese people. Hitherto, these have tended to be treated in isolation from each other. There are books on Japanese technology such as Erich Pauer's edited collection of papers entitled *Papers on the History of Industry and Technology of Japan*, 3 vols (Marburg, Germany: Förderverein "Marburger Japan-Reihe," 1995), in which each paper tends to be discrete and focuses on the technology itself. In *The Technological Transformation of Japan* (Cambridge: CUP, 1994), Tessa Morris-Suzuki weaves her case studies into a more coherent whole, using the concept of social networks to help explain the transition from small-scale factories and workshops to large modern enterprises. This book builds on her work and her interest in social concerns, but rather than limiting itself to technology, we also look at the contribution of science and medicine in helping to shape modern Japan and the very bodies of the people themselves. There are books such as William Johnston's *The Modern Epidemic: A History of Tuberculosis in Japan* (Cambridge, Mass.: Council on East Asian Studies, Harvard University, 1995), but unlike that volume, we examine a variety of diseases and ailments, and how the state sought to deal with them.

Versions of some of the chapters included in this book were presented at annual meetings of the Association for Asian Studies (AAS). We are grateful to the AAS for providing a forum for our ideas. The contributors to this book also acknowledge the support of their respective institutions. This book was completed while the editor was Senior Lecturer in Asian Studies at the University of Queensland.

Following normal East Asian practice, Japanese and Chinese surnames precede given names, except for those authors who choose to use the Western order when writing in English. Macrons have been omitted from well-known place names such as Tokyo. In chapter 5, the term "Manchuria" is used to refer to the northeast China provinces of Heilongjiang, Jilin, and Fengtian, now referred to as Dongbei, the "Northeast." The Japanese name for Dairen is used instead of the city's Chinese name of Dalian.

Finally, we wish to express our warm thanks to Anthony Wahl, Heather Van Dusen, and Alan Bradshaw at Palgrave Macmillan, and Maran Elancheran and the copy editor at Newgen Imaging Systems who worked on this project. Without their interest, support, and patience, this book would not have been possible. Comments from an anonymous referee were also useful in revising the chapters for publication.

MORRIS LOW

NOTES

1. *The Economist*, "(Still) Made in Japan," April 10, 2004, 57–59 esp. p. 57.
2. Brendan Pearson, "Boom in Japanese Sales to Asia," *The Australian Financial Review*, March 26, 2004, 27.
3. James F. Hoge, "Preparing for Asia's Eclipse of the West," *The Australian Financial Review*, June 25, 2004, "Review" supplement, 10–11.
4. Tessa Morris-Suzuki, *Re-Inventing Japan: Time, Space, Nation* (Armonk, NY: M.E. Sharpe, 1998), pp. 162–167.

Introduction

Morris Low

Commodore Matthew C. Perry's expedition to Japan in 1853 and 1854 is characterized as having "opened up" Japan[1] and ushered in a period of transformation beginning with the Meiji Restoration (1867–1868). The history of the Meiji era (1868–1912) has received special attention[2] but the role of science, technology, and medicine in the transformations that Japan underwent at that time and in the decades that immediately followed, has yet to be revisited.

Perry's visits to Japan heralded a period of much greater interaction with the West. While the *Narrative of the Expedition*[3] and subsequent histories tend to depict the introduction of Western science and technology in terms of the importance of the artifacts that were exchanged,[4] namely a miniature steam engine and two telegraph sets, the real significance of the expedition for us is in how the gifts signaled Japan's entry into a Western-dominated, global system of knowledge in which the railway and the telegraph helped transmit information. Indeed, the telegraph has been likened to today's Internet.[5] It is thus possible to view the history of modern Japanese science, technology, and medicine in terms of discourse, ideas, and know-how. This is the crux of this book.

The art historian Timon Screech has written insightfully of the years before the Meiji era. In the eighteenth century, Western ideas and things came courtesy of the Dutch via their trading post at the port of Nagasaki. "Holland" came to represent a type of discourse related to imported objects such as telescopes, microscopes, spectacles, and kaleidoscopes. These things helped the Japanese understand the need for precision, and to see the world in new ways.[6]

The bodies of the Japanese people became increasingly technologized, "fitting into and effortlessly moving through new technological networks."[7] A famous woodblock print by Utagawa Yoshitora of the Tsukiji silk factory in Tokyo in the early 1870s shows kimono-clad female textile workers almost entrapped within an imported silk reeling machine.[8] Japan's modernization harnessed human labor for the purposes of industrial progress. The body also became the site for biological reconstruction and intervention. As we shall see, Meiji intellectuals became interested in eugenic ideas and race improvement theory.

During the early years of the Meiji period, the slogans *fukoku kyōhei* (a wealthy nation and a strong army), *shokusan kōgyō* (encouraging industry) and *bunmei kaika* (civilization and enlightenment) were important ways of promoting national policy and encouraging the Japanese people to contribute toward the good of the nation. The ideological dimension has been ably discussed by historians such as Carol Gluck[9] and Takashi Fujitani,[10] and James Bartholomew has written the key book on the formation of a scientific community at this time.[11] This book builds on their scholarship.

Despite the sacrifices made by the Japanese people, the elder statesman of Japanese politics Ōkuma Shigenobu attributed Japan's rapid progress in the years that followed to foreign intercourse.

> This leap forward is the result of the stimulus which the country received on coming into contact with the civilization of Europe and America, and may well, in its broad sense, be regarded as a boon conferred by foreign intercourse. Foreign intercourse it was that animated the national consciousness of our people, who under the feudal system lived localized and disunited, and foreign discourse it is that has enabled Japan to stand up as a world-Power. We possess to-day a powerful army and navy, but it was after Western models that we laid their foundations . . . . All this is nothing but the result of adopting the superior features of Western institutions.[12]

In a way, this book seeks to understand Japan's process of modernization by unpacking Ōkuma's words.

Commodore Matthew C. Perry's visits to Japan in 1853 and 1854 resulted in the signing of the U.S.–Japan Treaty of Amity and Commerce in July 1858. This treaty, and other similar treaties signed with Holland, Russia, Britain, and France, effectively opened Edo (Tokyo), Osaka, Kanagawa (Yokohama), Nagasaki, Niigata, and Hyōgo (Kobe) to trade with the West in subsequent years.[13] These "unequal treaties" stipulated extraterritoriality for foreigners in Japan whereby foreigners accused of crimes would be tried in courts by foreign judges under foreign laws. The treaties also gave the Western nations favored-nation treatment in trade. The agreements thus set the rules for "foreign intercourse," albeit skewed in favor of the West.

The Meiji Restoration (1867–1868) was not unlike a *coup d'état* that "restored" the young Emperor Meiji to sovereign power. It heralded the transfer of political power from one group of samurai to another and ushered in a program of modernization that involved not just opening Japan to trade with the West but participating in a global system of knowledge—"foreign intercourse" in the broadest sense of the term.

As Ōkuma suggested, Japan did go shopping for Western organizational models that were adopted and adapted along with manufacturing and communications technologies.[14] These changes transformed Japanese society and what emerged were sometimes hybrid forms that drew on both foreign and the local. Western institutions such as schools and universities, the army and the navy, and systems of central and local administration provided the framework through which the cultural heritage of Japan could be translated. The next 50 years of rapid modernization, involved a series of complex interactions with things Western by a multitude of Japanese. Their common goal was to build a modern Japan. This was achieved not only by borrowing from the West but also by considerable innovation and actively building on existing resources.

In order for Japan to communicate and interact with the rest of the world, Japan needed to translate indigenous knowledge and expertise into internationally accepted forms of knowledge and behavior.[15] The first part of this book examines how scientific and medical discourse was shaped and applied, a sense of nation engendered, and loyal subjects created. The lives of the Japanese people came increasingly under the control of the state, and there were attempts to extend such control to colonial territories, in the hope of maintaining a healthy nation and empire. The second half of the book is characterized by careful studies of the adoption of Western technology, state–business interaction, and the growth of industries. We see that as in science, much of the history of technology is in fact about ideas, information, and know-how, and how they are applied and used.[16] Successful technology transfer in Meiji Japan was arguably more about careful on-the-job training and adaptation of Western techniques rather than mere importation of the relevant piece of equipment.

We begin with the health of the nation. As Christian Oberländer outlines in chapter 1, it was in the years after the Meiji Restoration that public health became an area of government concern and subject to state control. Medicine was expected to contribute to making Japan a "rich country with a strong army." The adoption of Western medicine was very much a part of Japan's modernization, leading to the introduction of scientific medicine, which involved the establishment of institutions, namely the hospital and the laboratory, both of which involved research.

The prevalence of diseases such as smallpox and beriberi, as well as cholera epidemics, spurred the government into taking action.[17] Beriberi, which at the time was considered infectious, garnered government attention from the late 1870s, as it was becoming a problem in the army and the navy.

Oberländer throws light on the process by which the Japanese adopted scientific medicine by focusing on the search for the cause of beriberi. Like many of the other histories contained in this book, the story is far from straightforward. It involved the opening of a state Beriberi Hospital which represented Japan's first major modern medical research program. We also learn of the transfer of bacteriological techniques from Germany to Japan, and competing claims by different researchers.

Oberländer effectively shows us how Japan bought into a global knowledge system where there were commonly shared ideas and techniques. Medical scientists working in laboratories tested their hypotheses through controlled experiments, and Japanese researchers were increasingly able to participate as members of the international scientific community.[18]

In chapter 2, Sabine Frühstück writes of another ailment that afflicted the Japanese (especially men): neurasthenia. There was speculation that it was linked to certain sexual practices, namely sexual abstinence, masturbation, or homosexuality. Like beriberi, neurasthenia was considered as threatening the health and well-being of the nation. Its occurrence in Japan served to reassure the Japanese in an odd way that Japan, too, suffered from afflictions common to other parts of the developed world. Japan was now part of the Western world order and suffering accordingly! Frühstück's key point is that the emergence of sexological discourse in these countries gave rise to neurasthenia. Scientific theories emerged to help explain such phenomena, with concepts such as "nerve force" being popular. These theories and the language of the discourse, linked disease to modernity and progress.

Physical and mental weaknesses were seen as impairing the advancement of the nation, with serious consequences for the political, military, and economic strength of Japan itself. Medical discourse thus sought to mould sexual practices toward the good of the nation. They provided frameworks, which helped define gender and sexuality.

In chapters 3 and 4 respectively, Sumiko Otsubo and Yuki Terazawa relate how there were attempts to medicalize everyday life and influence the reproductive lives of the Japanese people. The introduction of the concept of race as a scientifically valid category paved the way for Social Darwinism to take root in Japan. Otsubo's chapter deals with the introduction of eugenic thought into Japan in the late nineteenth and early twentieth centuries. Her chapter provides us with another example of the selective adaptation of ideas and institutions that occurred in the Meiji era. Otsubo focuses on the ideas of the Tokyo University professor of physiology Ōsawa Kenji and his role in the spread of eugenics and race improvement theory in Japan. Ōsawa saw the female body as being an important site for the improvement of the Japanese race. He medicalized the institution of marriage by encouraging the issuing of prenuptial health certificates, and promoted careful eugenic control of marriage that is, selective breeding. Through such policies, the state sought to influence the private lives of Japanese citizens.

In a related paper, Yuki Terazawa focuses on the physician Yamazaki Masashige to argue how scientific and medical discourses were developed in conjunction with other discourses relating to Japan's nation-building and empire-building agenda. Some Japanese intellectuals at the time were reconciled to the idea that the Japanese were inferior to Western people but they took consolation in the belief that by creating a healthy environment (not only in terms of hygiene and education but also socially and economically) the physical and mental capacity of citizens would be improved to a point where Japan would be on par with the West.

Terazawa situates Yamazaki's ideas within the context of anthropological studies of racial difference. She examines Yamazaki's work on the menstruation of women of different racial backgrounds: Japanese, Ainu, Ryūkyūan, and Chinese. Most interestingly for our purposes, Yamazaki emphasized how such studies of women's reproductive capacity were important for the nation. Yamazaki sought to correlate the reproductive physiology of different races to the degree of "progress" each group had attained. Not surprisingly, he used this to argue that the Japanese were superior to others in the Japanese empire. In the process, Yamazaki's research helped to establish "the Japanese" as a unitary group, a nation whose borders and territorial expansion were justified racially and scientifically.

The chapters by Frühstück, Otsubo, and Terazawa illustrate how discourses served to structure not only the Japanese sense of reality but also their very notion of identity and place within the world. While the ideas that all three authors discuss are fascinating in their own right, it is the response to these ideas and concepts, the way that they are adapted, and the practices that arise from them, that are especially cause for concern.

Ōkuma Shigenobu's own ideas about Japan's progress, mentioned earlier, were informed by Social Darwinism. By "adopting the superior features of Western civilization,"[19] the Japanese race could evolve and Japan could be transformed into a great nation. By the 1920s and 1930s, such ideas helped popularize the practice of eugenics—government regulation of human reproduction. There was a perception that the Japanese race still needed improvement if Japan was to be the equal of Western nations. This culminated in the National Eugenics Law of 1941, which sought to prevent people with serious hereditary diseases from reproducing.[20] Such policies were more difficult to implement away from Japan proper.

Chapter 5 helps bridge the two parts of this book. Japan's project of modernization was exported to its colonies and territories under its control. As Tessa Morris-Suzuki has pointed out, turning colonial subjects into "Japanese" required the state to intervene in their lives far more intrusively than in the home islands of Japan.[21]

There was a belief that building a modern nation involved keeping the empire healthy. Robert Perrins shows how controlling disease in colonial territories overlapped with planning for economic growth and development. Japanese authorities in southern Manchuria sought to improve public health by a variety of means including city planning in order to protect their investment, to showcase the modernity that accompanied Japanese colonial rule, and to create a healthy enclave suitable for the growing Japanese population.

Why do certain technologies flourish in some places and not elsewhere? The second part of the book helps to answer such questions. One of the factors in Japan's modernization has been the significant role of state–private sector relations. As Richard J. Samuels points out, a type of techno-nationalism emerged that linked nation-building with technological development (*gijutsu rikkoku*).[22] Samuels has written of how the Meiji program of industrial development involved the protection of existing industries, subsidies to nurture industries, the introduction of new machinery, and the transfer of state-owned model factories to the private sector. Particularly important was the textile industry, which David Wittner focuses on in chapter 6.

Wittner examines the role of the government in the early Meiji period in terms of its assistance to the silk industry. Most of the later expansion in the textile trade was not the result of government policy. The story of Tomioka suggests why. Government-led industrialization was not a coherent program of initiatives and ventures. Rather, Wittner shows that it was rather *ad hoc* in nature with often little detailed planning and deliberation involved.

In the rush to obtain lessons from Meiji Japan's success, there is a tendency to view Japan as a monolithic whole, an entity whose actions resulted in a series of successes rather than failures. This is far from the truth. Wittner focuses on the mechanization of Japan's silk industry. The hiring of the Frenchman Paul Brunat as chief silk reeling adviser shows how recruitment of foreign employees was somewhat haphazard. Personal endorsements and perceptions of trustworthiness seemed more important than qualifications. The way in which the type of technology was chosen appears problematic as well, with Brunat opting for hybrid reeling techniques that incorporated local technology.

Wittner makes the startling point that the choice of technique and the technology transfer was determined more by ideology, than technical or economic considerations. Politics at both a local and international level were very important. The choice of French silk reeling technology at the model filature at Tomioka helped legitimate the new government and reflected the seriousness of intent. But even within Japan, doubts were voiced as to whether or not the technology, given its cost and complexity, was really appropriate for Japan and whether Brunat was sufficiently qualified to lead the project. The idea of an all-knowing, rationalist Meiji state falls apart. Chapter 7 provides further reasons for why we need to question such assumptions.

Western buildings in the treaty ports required glass windows. In chapter 7, Chaiklin explains why it took so long for the domestic production of sheet glass to occur in Japan. In the meantime, the demand was largely met by imported glass. It was not until 1910 that commercial production began in earnest, and a further ten years before production commenced at the American Japan Glass Company, a successful joint venture with the Libbey-Owens Sheet Glass Company. Not only did the business arrangement facilitate the borrowing of a mechanized rolling method to produce sheet glass, but it also ensured that Japanese artisans were properly trained and supervised over an extended period of time.

The industry had a checkered history of financial losses by both the government and private entrepreneurs that belies the term "miracle" that is so often used to describe Japanese economic growth. But motives were not only financial. The need to show that Japan was "civilized" and the equal of Western nations meant that some of the changes that Japanese architecture underwent were initially mere window dressing.

Technological change had wide-ranging implications for the building trade. In chapter 8, Gregory Clancey focuses on the role of *daiku*, Japanese carpenters, and how they fared during this time of considerable upheaval. Clancey, like Chaiklin, warns us about simplistic narratives of progress. For Clancey, carpenters served a variety of roles, from facilitators of Westernization to a threatened group of craftsmen. Like other Japanese in "traditional" occupations (such as practitioners of Chinese medicine), the Meiji period was a time of shifting occupational definitions, a struggle to survive and adapt with the times and its changed expectations. *Daiku* achieved positions of authority in the Ministry of Public Works from early on, even before Western-style architects came to be trained by the Ministry in the 1870s.

In a way, *daiku* bridged both public and private, and Japanese and Western ways. Clancey refers to how they developed *wayō setchū* (Japanese–Western compromise) architecture, which combined elements from both cultures. *Daiku* geometry was used to "translate" Western carpentry forms rather than *vice versa*. Chaiklin relates how glass windows were able to be incorporated into traditional architecture, and how the changes met with little cultural resistance. Clancey's chapter, and the chapters by Wittner and Chaiklin, all serve to remind us that Western technology was not introduced into a vacuum. There were local knowledge systems in place and modernization was a process of negotiations and compromises. Rather than portraying the world of the *daiku* in terms of buildings and tools,[23] Clancey sees it as a discursive world where their sense of identity was shaped and redefined.

The decades that followed saw considerable negotiation between public and private actors as to where the role of state institutions stopped in terms of the economy. In chapter 9, W. Miles Fletcher helps us to understand business–government relations in these later decades through a case study of the cotton spinning industry. It was the first large-scale mechanized industry in Japan and the first to become internationally competitive. How can we account for its success? The chapters by Wittner and Chaiklin show how we need to be wary of the notion of the "all-knowing" Meiji state. Similarly, in Fletcher's chapter, the policies of the Japanese government in the 1920s and 1930s certainly appear to have been *ad hoc* and at times deeply flawed, especially in officials' slow response to the onset of the Great Depression in 1930. The Japanese spinning sector, Fletcher shows, feared the clumsy hand of governmental interference and strongly desired to protect its autonomy in managing its own affairs.

The cotton textile industry quickly became part of the global world order that Japan had signed up for in the nineteenth century. Overseas sales of cotton goods began in the 1890s and by the 1920s was second only to raw silk in terms of contributing to the nation's exports. From the start, because spinning firms became dependent on imported raw cotton and overseas markets, business leaders had to pay close attention to developments abroad. Just as Japanese entrepreneurs in the Meiji era had to cope rapidly and adroitly with the challenge of adapting to unfamiliar Western technology, in the 1920s, executives had to adjust to a new set of international challenges, first in the form of rising trade barriers in important markets and then the worldwide dampening of demand by the Great Depression.

The Japan Spinners Association had played a crucial role in the growth of the spinning sector by promoting close cooperation among cotton textile firms. In crises, the association followed an intricate strategy of modulating output to balance supply and demand while encouraging the expansion of production capacity and the installation of more efficient equipment. Its monthly journal publicized new technology and transmitted best practice among firms.[24] Fletcher argues that although the Great Depression wielded a blow to the industry the Spinners Association's effective response merely reaffirmed the tradition of self-governance. Far from turning to the government for aid, the spinners used their own ingenuity to devise policies for overcoming the depression within a year. When proliferating trade barriers—a development well beyond the industry's direct control—forced the government to regulate exports to some markets, the association fought successfully to maintain its influence in the determination of national trade policy.

Fletcher's chapter also provides important insights into the dramatic changes that Japan's domestic economy underwent during the 1930s. Many scholars, such as Takatoshi Ito, agree that the Depression did not have as severe an impact in Japan as it did in the United States and Europe. One explanation is the huge increase in the government's deficit spending and in military expenditures after the Manchurian Incident of 1931. Faced with new trade barriers overseas, Japan, like other major powers, turned to its empire to secure both sources of raw materials and export markets. This strategy served to accelerate Japan's military expansion.[25] But Japan's recovery depended on much more than military spending and colonies. As Andrew Gordon points out, Japanese industrial production grew an astonishingly high 82 percent by 1934. Exports outside of the empire doubled in the years from 1930 to 1936, and Japan rose to become the leading exporter of cotton goods.[26]

The conflict with China in July 1937 ushered in direct government control of industries in the name of the war effort. And even when such controls were lacking, the needs of the military received priority. For example, in the flat glass industry, there were, by 1937, a total of 5 companies in Japan with 8 plants and 14 furnaces. Total monthly production is said to have ranked among the highest in the world. However, in the years leading up to the Pacific War in 1941, the output of sheet glass declined and production shifted to polished plate glass and safety glass for use in military vehicles.[27] The manufacture of cotton goods plunged after 1941, and many spinning factories were forced to switch to the production of munitions. Of the 11,434,816 spindles operating in 1941, only 2,184,122 (19.1%) were in use at the end of the war in 1945.[28] Other industries experienced similar changes.

In many ways, the central developments examined in this book—the absorption of global scientific knowledge, the strengthening of the nation through improved public health, the adaptation of Western technology, successful industrialization, and effective responses to economic crises—made possible Japan's ambitious war to create a Greater East Asia Co-Prosperity Sphere. Industrial power enabled the nation to field an army and a navy that would conquer much of China and challenge the Anglo-American powers. At the start of the Pacific War, some Japanese military technology was clearly superior to that of the most advanced Western powers. Long-range torpedoes and the Mitsubishi "Zero" fighter are among the best known examples. Paradoxically, the scale of the devastation and suffering caused by the war threatened to bring Japan's progress to a halt. By the time of the Japanese surrender, extensive Allied bombing had reduced the nation's major cities to rubble and had destroyed nearly one-half of its factories. Strict rationing of food had brought many Japanese close to starvation. Yet, the knowledge, skills, intellectual curiosity, and sheer determination that had driven the nation's achievements in the early twentieth century proved durable. Japan quickly rebuilt again, with vigor.

As we have found in this volume, it was not so much the material infrastructure that was put into place but the Japanese people who were key to Japan's modernization. As they became an increasingly important part of the global world order, actively participated in international trade, and became part of the global knowledge system, the bodies of the Japanese people became "intermingled with machines, . . . pierced by information, and physically transformed by ideas."[29] Ōkuma Shigenobu was right in attributing Japan's rapid progress to foreign intercourse, rather than the West per se. It was through interaction with the West, and a multitude of encounters with its ideas and discourses, that Japan negotiated its way to becoming a modern and prosperous nation.

NOTES

1. See, e.g., Peter Booth Wiley with Korogi Ichiro, *Yankees in the Land of the Gods: Commodore Perry and the Opening of Japan* (New York: Penguin Books, 1991).
2. Helen Hardacre, with Adam L. Kern, eds., *New Directions in the Study of Meiji Japan* (Leiden: Brill, 1997).
3. Francis L. Hawks, *Narrative of the Expedition of an American Squadron to the China Seas and Japan, Performed in the Years 1852, 1853, and 1854, under the Command of Commodore M.C. Perry, United States Navy, by Order of the Government of the United States* (New York: Appleton, 1856).
4. Chang-su Houchins, *Artifacts of Diplomacy: Smithsonian Collections from Commodore Matthew Perry's Japan Expedition* (1853–1854) (Washington, D.C.: Smithsonian Institution Press, 1995), p. 149.
5. Tom Standage, *The Victorian Internet: The Remarkable Story of the Telegraph and the Nineteenth Century's Online Pioneers* (London: Phoenix, 1999).
6. Timon Screech, *The Lens within the Heart: The Western Scientific Gaze and Popular Imagery in Later Edo Japan*, second edition (Honolulu: University of Hawaii Press, 2002), Chapter 1.
7. Iwan Rhys Morus, "The Measure of Man: Technologizing the Victorian Body," *History of Science* 1999, *37*: 249–282, on p. 249.
8. See Tessa Morris-Suzuki, *The Technological Transformation of Japan: From the Seventeenth to the Twenty-first Century* (Cambridge: Cambridge University Press, 1994), cover illustration and Figure 4.1, p. 76.
9. Carol Gluck, *Japan's Modern Myths: Ideology in the Late Meiji Period* (Princeton: Princeton University Press, 1985).
10. Takashi Fujitani, *Splendid Monarchy: Power and Pageantry in Modern Japan* (Berkeley: University of California Press, 1996).
11. James R. Bartholomew, *The Formation of Science in Japan: Building a Research Tradition* (New Haven: Yale University Press, 1989).

12. Shigenobu Ōkuma, "Conclusion," in *Fifty Years of New Japan (Kaikoku Gojūnen Shi)*, ed. Shigenobu Ōkuma and Marcus B. Huish (London: Smith, Elder, and Co., 1909), pp. 554–575, esp. p. 555.
13. W.G. Beasley, *The Meiji Restoration* (Stanford: Stanford University Press, 1972), p. 108; Janet Hunter (comp.), *Concise Dictionary of Modern Japanese History* (Berkeley: University of California Press, 1984), pp. 233, 240.
14. D. Eleanor Westney, *Imitation and Innovation: The Transfer of Western Organizational Patterns to Meiji Japan* (Cambridge, Mass.: Harvard University Press, 1987).
15. Tessa Morris-Suzuki, "The Great Translation: Traditional and Modern Science in Japan's Industrialisation," *Historia Scientiarum* 1995, *5* (2): 103–116; Tessa Morris-Suzuki, *Re-Inventing Japan: Time, Space, Nation* (Armonk, N.Y.: M.E. Sharpe, 1998), p. 163.
16. Hans Christian von Baeyer, *Information: The New Language of Science* (London: Weidenfeld and Nicolson, 2003).
17. William Johnston, *The Modern Epidemic: A History of Tuberculosis in Japan* (Cambridge, Mass.: Council on East Asian Studies, Harvard University, 1995).
18. Tessa Morris-Suzuki, Re-Inventing Japan, p. 165.
19. Shigenobu Ōkuma, "Conclusion," pp. 554–575, esp. p. 555.
20. Sumiko Otsubo and James R. Bartholomew, "Eugenics in Japan: Some Ironies of Modernity, 1883–1945," *Science in Context* 1998, *11* (3–4): 545–565, on p. 546.
21. Tessa Morris-Suzuki, Re-Inventing Japan, p. 170.
22. Richard J. Samuels, *"Rich Nation, Strong Army": National Security and the Technological Transformation of Japan* (Ithaca: Cornell University Press, 1994), p. 42.
23. William H. Coaldrake, *The Way of the Carpenter: Tools and Japanese Architecture* (New York: Weatherhill, 1990).
24. Gary R. Saxonhouse, "Country Girls and Communication among Competitors in the Japanese Cotton-Spinning Industry," in *Japanese Industrialization and Its Social Consequences*, ed. Hugh Patrick with the assistance of Larry Meissner (Berkeley: University of California Press, 1976), pp. 97–125, esp. 122–123.
25. Takatoshi Ito, *The Japanese Economy* (Cambridge, Mass.: The MIT Press, 1992), p. 14.
26. Andrew Gordon, *A Modern History of Japan: From Tokugawa Times to the Present* (New York: Oxford University Press, 2003), p. 192.
27. Fujimura Hiroshi, Hirao Genyū, Naitō Masao, and Sakata Hironobu, "The Development of the Flat Glass Industry in Japan," in *The Development of the Japanese Glass Industry: Papers on the History of Industry and Technology of Japan, Vol.3*, ed. Erich Pauer and Sakata Hironobu (Marburg: Marburger Japan-Reihe, 1995), pp. 45–64, on p. 49.
28. Keizo Seki, *The Cotton Industry of Japan* (Tokyo: Japan Society for the Promotion of Science, 1956), p. 311.
29. Ollivier Dyens, trans. Evan J. Bibbee and Ollivier Dyens, *Metal and Flesh, the Evolution of Man: Technology Takes Over* (Cambridge, Mass.: MIT Press, 2001), p. 1.

Part 1

Science, Medicine, and a Healthy Nation

1

The Rise of Western "Scientific Medicine" in Japan: Bacteriology and Beriberi

Christian Oberländer

Introduction

The adoption of Western medicine was an integral part of Japan's modernization from its very beginning,[1] leading ultimately to the introduction of "scientific medicine," a defining characteristic of the modern world. Scientific medicine started as a development in Western Europe, and after considerable conflict, came to be recognized as producing "true medical knowledge." This, in turn, was made universal through exportation.[2] Scientific medicine is based on two distinctive institutions, the hospital and the laboratory, which still prevail today. In hospital medicine, clinical investigation searches for correlations between symptoms and signs of disease, and internal changes of the body. Research focuses on anatomical pathology, and postmortems are routinely performed. In laboratory medicine, causes of diseases are identified by experiments in order to create cures for them. Laboratory research concentrates on living processes like bacteriology, uses living animals for experiments, and depends strongly on scientific instruments like microscopes.[3]

Historical research on Western medicine in Japan so far has paid little attention to the process of adopting scientific medicine. However, the investigations that the Japanese authorities and individual physicians carried out during the early Meiji period to identify the cause of beriberi (in Japanese, *kakke*), temporarily culminating in Ōgata Masanori's[4] (1855–1919) discovery of a "beriberi bacillus" in 1885, present an important case through which we can glean insights from the Japanese experience into the rise of scientific medicine in a non-European society. Because beriberi was not prevalent in Europe when it became a public health challenge in Japan, there were no ready-made containment policies available and the Japanese government had to try controlling this menace single-handed. As Japan had managed to escape the threat of colonization and could develop quite autonomously, early beriberi research allows us to examine the introduction of scientific medicine in Japan under "laboratory conditions."

Japanese authors who discuss the history of beriberi in Japan dismiss bacteriological research on this disease, and Ōgata's work in particular, as only a diversion from the "true" path of medical progress that eventually led to the discovery of the cause, deficiency of vitamin B_1.[5] The historian of Japanese bacteriology Fujino Tsunezaburō explains Ōgata's work in detail, but does not place it in the context of Japanese beriberi research.[6] In European languages, Ōgata's discovery has been treated almost exclusively with regard to his later dispute with Japan's internationally more famous bacteriologist, Kitasato Shibasaburō (1852–1931). The alleged consequences that Kitasato's critique of Ōgata's findings had for Kitasato's career have been at the center of attention, rather than Ōgata's discovery itself.[7] K. Cordell Carter, too, in his study of the history of beriberi research, does not consider Ōgata's work in detail.[8] This chapter seeks to do so by asking in particular how scientific medicine was adopted in the course of early beriberi research during Japan's early modernization, and what role the germ theory played in this process.

Beriberi as a Threat to Modernization

After the Meiji Restoration (1867–1868), the role of medicine in Japanese society changed dramatically. Medicine was now expected to contribute to the government's policy of modernization, symbolized by the slogan of a "rich country with a strong army" (*fukoku kyōhei*). This shift had already begun under Tokugawa rule at the end of the Edo period, but after the Meiji Restoration, the new government strongly promoted this process. In 1872, an "Office for Medical Affairs" (*Imu-ka*) was created within the Ministry of Education. This was later succeeded by the "Bureau of Hygiene" (*Eisei-kyoku*) under the leadership of Nagayo Sensai (1838–1902) of the powerful Ministry of the Interior. The Bureau was responsible for regulating health care and coordinating the numerous measures that were now taken to defend the public's health. In 1874, the "Medical Act" (*Isei*) established the first national licensing examination for physicians based on Western medicine, effectively abolishing traditional Kanpō medicine.[9]

During this early stage of the creation of a medical administration, beriberi was of little concern. Epidemic diseases such as cholera that threatened almost the entire country captured the attention of the health authorities. However, beriberi's prevalence had risen already during the late Edo period and it had become a permanently present endemic disease. From the great population centers—Edo (later to become Tokyo), Osaka, and Kyoto—the disease had spread to provincial towns and then to rural areas. The thirteenth and the fourteenth shoguns, Tokugawa Iesada (1824–1858) and Tokugawa Iemochi (1846–1866), are said to have both died of beriberi at the ages of 34 and 20, respectively.[10]

Beriberi suddenly moved to the center of attention when during the Seinan civil war of 1877, a large percentage of the troops fell sick with this disease. According to contemporary statistics, the rate of affliction with beriberi in the Japanese army had been 11 percent in 1876, but climbed to 14 percent in 1877, and jumped to 38 percent in 1878.[11] In the Japanese navy, the prevalence was 33 percent in 1878. If beriberi damaged the fighting capabilities of Japan's troops during a civil war this heavily, how much more crippling would the effects of the disease be during a military confrontation on the Asian continent?[12] The Japanese government decided that, in order to protect the combat readiness of the Japanese armed forces, the beriberi problem had to be solved.

However, there was a second important reason why the medical administration became increasingly concerned with beriberi at this particular time. Since April 1876, the Japanese empress suffered gravely from the disease. One year later, she recovered, but then in June 1877, Kazunomiya, the emperor's sister, fell seriously ill with beriberi, and one month later, even the emperor himself contracted the disease. When in September 1877, Kazunomiya died from heart failure caused by beriberi, this was a severe shock to the Japanese imperial family. The trust in Western medicine that had replaced Chinese-style medicine at the imperial court since the Meiji Restoration was deeply shaken and some physicians of traditional Kanpō medicine were reappointed to court offices. While the emperor recovered by the end of the year, he was now sensitized to the problem of beriberi, and his continuing interest in combating beriberi is reflected in numerous recordings in the imperial chronicle.[13] This was especially so since the damage done by the disease was not limited to the imperial family, but every year, there were thousands of victims among the civilian population. Beriberi thus put the supreme aim of Japanese modernization—to become a "rich country with a strong army" in order to remain an independent nation—in doubt.

As the first official measure of research into the cause of beriberi, in December 1877, the Bureau of Hygiene started a national initiative to collect and evaluate all knowledge about the disease available in Japan. In order to include even the most recent observations, Ministry of the Interior officials instructed all public hospitals throughout the country to gather all extant information concerning the pathology and treatment of the disease and to submit it to the Bureau during January 1878. In the internal explanation for this order, the officials of the Ministry of the Interior stressed that beriberi was confined almost exclusively to Asia. Even the foreign physicians who were practicing in different prefectures of Japan since the Meiji Restoration, despite having conducted numerous investigations of the pathology and therapy of the disease and having formulated different theories regarding its cause, had no confirmed insights in either area of inquiry. Therefore, the Bureau of Hygiene wanted to carry out a comparative study of the different academic theories and practical approaches.[14]

The directive of the Ministry of the Interior of December 1877 caused a wave of concern regarding beriberi among Japan's medical community, and the number of monographs published on the subject rose steeply in 1878. The physicians of Kanpō medicine reacted especially quickly because they viewed it as an opportunity to win support for Chinese-style medicine that was being abolished. Kanpō physicians submitted petitions to the government outlining their treatment methods and claiming that their medicine was equipped with more effective cures for beriberi than Western medicine because they had more experience treating this disease that was prevalent mainly in Asia. In addition to offering their know-how, Kanpō physicians also founded private beriberi hospitals to demonstrate the special effectiveness of Kanpō medicine to government officials and to gain sympathy for Chinese-style medicine among the common people.[15]

An example of the theories drafted by Kanpō practitioners provides the "New Treatise of Beriberi Disease" (*Kakke shinron*) that was published in May 1878 by the famous Kanpō doctor Imamura Ryōan (1814–1890) who had been a court physician of the Tokugawa family. Imamura's work was based on medical theories stemming from the Chinese Tang period. In China, one of the ways that physicians had explained the origins of the disease was the "theory of outer causes" (*gaiinsetsu*) based on a "wind poison" (*fūdoku*) that supposedly originates in the soil and enters the body through the legs. Imamura enriched this theory with his personal experience and in part even included anatomical concepts from Western medicine. He argued that a "poison" would enter the inner organs via the blood vessels and, as the disease progresses, spreads into the heart and lungs where it would cause the characteristic attacks.[16]

Besides the Kanpō doctors, the Japanese physicians practicing Western medicine now paid more attention to beriberi, and this was reflected in Japan's new medical journals.[17] In April 1878, for example, the *Tokyo Medical Journal* (*Tōkyō iji shinshi*), one of the most influential medical periodicals in Japan during the early Meiji period,[18] published a paper "On Beriberi" (*Kakke-ron*) written by the physician Kashimura Seitoku (1857–1902). Kashimura, who later became one of the research directors of the government's Beriberi Hospital, suspected the cause of the disease to be a "malaria poison" (*mararia doku*).[19] In this, Kashimura possibly followed the lead of the prominent army surgeon Hashimoto Tsunatsune (1845–1909) who in 1876 had submitted to the University of Würzburg a German-language dissertation "About the Beriberi Disease," a Japanese language summary of which was printed in the first edition of the Medical Newspaper (*Iji shinbun*) of May 1878. Concerning the origin of beriberi, Hashimoto had quoted different theories, for example the assumption of an inflammation or softening of the spinal cord, but personally he favored a "miasma formation" in swamps as the most likely cause of the "malaria-like" illness.[20] Another medical officer of the army, Ishiguro Tadanori (1845–1941), authored his own "Theory of Beriberi" in August 1878. According to Ishiguro, a "fungus"[21] caused the disease. This "fungus" supposedly formed as the result of transformation processes in the polluted soil of the great population centers, moved from the ground into the atmosphere and entered the human body through drinking water. As a causal therapy, he recommended fighting the "fungus" with quinine.[22]

In the late 1870s, beriberi was increasingly perceived as a threat to Japan's modernization policy. In addressing this threat, Japan was faced with two-fold difficulties. First, in contrast to other diseases such as cholera against which the counterstrategies of Western countries could serve as a ready reference for Japan, beriberi was a disease that was little known in Europe, and Japan had to develop adequate preventive measures by itself. Second, knowledge regarding the causation and treatment of beriberi still either centered on the ancient theories of Chinese medicine or simply echoed versions of the then popular concept of "miasma" proposed by many European physicians.[23] Therefore, the officials at the Japanese Ministry of the Interior decided to make beriberi research a matter of state responsibility.

STATE-SPONSORED HOSPITAL MEDICINE AT THE BERIBERI HOSPITAL

The national survey of knowledge about beriberi led the officials at Japan's Ministry of the Interior in February 1878 to the conclusion that neither Kanpō nor European medicine provided a satisfactory theory of causation, not to mention an effective method of treatment of beriberi. In a memorandum to the State Council (*Dajōkan*), they requested funds to found a specialized hospital under the direction of the Bureau of Hygiene that would conduct comparative research on beriberi. The Bureau declared that the strengths and weaknesses of Kanpō and European medicine should be compared on the basis of their clinical performance and the origin of the disease should be elucidated through basic research. In spite of the high costs involved at a time of great fiscal strain, three days later, the State Council responded positively to the Ministry's request. The Imperial court, being afflicted heavily by beriberi, participated not only in generously financing the Beriberi Hospital, but also demanded that Kanpō medicine would be included in the trials.[24]

By July 1878, the state Beriberi Hospital opened its doors. The four clinical wards were placed under the authority of four leading physicians. Two of them, Tōda Chōan (1819–1889) and Imamura Ryōan, were representatives of Kanpō medicine, while the other two, Kobayashi Tan (1847–1894) and Sasaki Tōyō (1838–1918), practiced European style medicine. The effectiveness of their treatment plans was to be compared systematically. Ikeda Kensai (1841–1918) and Miyake Hiizu (1848–1938)—leading authorities on internal medicine and pathology—were responsible for basic research.[25]

The clinical side of the state Beriberi Hospital exemplified the multitude of causative theories and therapeutic approaches that were common in Japan at the end of the 1870s. On the part of the Kanpō physicians, Imamura treated his patients according to the established recipes of Chinese-style medicine. Tōda, a former court physician of the Tokugawa as well as the emperor, on the contrary abided by the secret method handed down within his family that was based on the conviction that the cause of beriberi was found in rice. He prohibited his patients from consuming rice and prescribed instead a diet based on Azuki beans.[26] Among the two physicians of European medicine, Kobayashi believed the precise cause of beriberi to be unknown but suspected a disease process similar to the ideas of Agathon Wernich (1843–1896)—a German lecturer for Internal Medicine at the university in Tokyo—who had pointed out various pathological signs of nutrition deficiencies that were supposedly caused by an inflammation of the digestive tract of beriberi patients. Kobayashi thus implemented a strict regimen of improved nutrition that required his patients to drink large quantities of milk. Sasaki who had studied European medicine with the Dutch military surgeon Johannes L.C. Pompe van Meerdevort (1829–1908), who was probably the first European to describe the Japanese "variation" of beriberi,[27] finally acted like a representative of Ishiguro whose theories and therapy suggestions he followed. Sasaki assumed a "fungus" to be at the root of beriberi and therefore treated his patients "causally" by administering quinine, otherwise addressing only the symptoms of the disease. Despite the variety of therapeutic approaches, there were, in the end, no significant differences between the curative successes of Kanpō and European medicine. Only Kobayashi achieved slightly better results than his colleagues with his nutrition-oriented milk therapy, while Sasaki stayed somewhat behind the field, perhaps because of his focus on fighting the suspected "fungus."[28]

How did basic research fare at the state Beriberi Hospital? While both of the two highly acclaimed research directors installed in 1878, Ikeda and Miyake, left the Beriberi Hospital after only a short period of time, the basic proceedings that they established were followed by their successors. From the outset, the scientific work was based on the assumption that beriberi was an infectious disease caused by a certain "poison" (*doku*) that entered the human body from the outside. The first and most pressing aim of the investigation was to observe the climatic and other circumstances under which the disease occurred in order to identify its "cause" (*gen'yū*) and pathology. If the factors leading to the formation of the "poison" and the mechanism of its entry into the human body were known, then a prevention of the disease would be possible, even if the exact "nature of the disease poison" (*byōdoku no honsei*) could not be fully understood. This pragmatic approach based not on a search for *the* cause, but the factors contributing to causation seemed justified to Ikeda and Miyake as they noted that even in Europe, the nature of many epidemic infectious diseases was not yet fully grasped. They wanted to investigate all circumstances that could possibly produce the disease. If they would remove that "factor which was closest" (*mottomo kin'in*) to the disease, then this should serve as a means of prevention.[29] As a second step, the afflicted organs and the most important symptoms of the disease were to be recorded according to precise clinical observation. The achievements of Kanpō medicine in this area were not considered sufficient as their focus was thought to be different from that of European medicine. Finally, an understanding of the pathology of the disease was Ikeda's and Miyake's third stated aim. Pathological dissections were to be conducted to observe the relationship between organic changes and the clinical course of the disease.[30] This research program at the Beriberi Hospital was thus characteristic of hospital medicine seeking to identify "factors" causing beriberi that could be linked to the clinical and pathological disease process and that could ultimately serve as a starting point for preventive measures.

The decision to base the scientific work of the Beriberi Hospital on the assumption of a disease poison followed the general trend at the time. In August 1880, the Medical Newspaper published a review of the beriberi research undertaken in Japan until then that included even the views of Pompe van Meerdevort and Antonius Bauduin (1822–1885) who had taught Dutch medicine in Japan during the Edo period (ca. 1600–1868). From this synopsis, it was clear that the majority of the leading Japanese representatives of Western medicine as well as their foreign colleagues believed a miasma to be the cause of the disease.[31]

The Japanese physicians perhaps accepted the notion of a miasma readily because the concept appeared similar to that of the "wind poison" of Chinese-style medicine of the Edo period. How did Japanese physicians of the early Meiji period imagine the miasma and its origin? In an essay on miasmatic disease of September 1880, the physician Satomi Giichirō explained that the exact nature of the miasma was not known, but it was thought that it was a kind of "mold" (*baishū*) that would enter the air with the evaporation of swamp water. In places where the swamps are shallow, the sunlight could reach the bottom and processes of decay would occur in the soil, causing the formation of the poison. Therefore, during hot summers, an especially large quantity of miasma was produced. When there was little air movement in a marshland area, the poison would remain near the swamps and would inflict harm only on the people in its immediate vicinity. Strong winds, however, would carry the miasma even to distant regions particularly threatening the lives of persons of younger age or of those who were weakened by another disease, especially when they were suffering from starvation. To support this theory, Satomi pointed to the example of malaria, which Max von Pettenkofer (1818–1901) claimed was caused by rising groundwater.[32]

During the following years, the researchers at the Beriberi Hospital investigated not only climatic influences but also age, sex, profession, and living conditions as factors of causation, as well as the amount and composition of the patients' urine. They also performed several postmortems.[33] In addition, they searched for explanations for why a change of location had been known since ancient times as the most promising method to treat beriberi and why foreigners were almost entirely spared by the disease. While they recognized the importance of nutrition as a predisposing factor,[34] they interpreted the changes in the spinal cord, nerve, and muscle tissue that were found in the pathological dissections only as symptoms of the disease and speculated that the "poison" was located in the blood. Studies of blood samples were meant to become the focus of research, but they could no longer be carried out as the Beriberi Hospital was closed in July 1882 after only four years of existence, and transformed into a special division within the university.[35]

As the first state-run large-scale medical research project with a modern program of inquiry typical of scientific medicine, the Beriberi Hospital went far beyond traditional approaches and was a novelty in Japan. It differed profoundly from other specialized beriberi hospitals based on localistic theories of disease. At the Japanese government's Beriberi Hospital, an ambitious program of hospital medicine was aiming for the discovery and confirmation of causative factors of beriberi. Patients served as a resource for medical research, and while postmortems did not readily become a matter of routine because of conflicting Japanese customs, they still formed a central part of the research program and were carried out by (foreign) experts with maximum circumspection.

The integrated research efforts did not, however, lead to the hoped-for breakthrough. The scientific results were meager compared with the huge financial investments and a starting point for preventive measures was not identified. The clinical observations, too, were not particularly startling because they largely replicated earlier work of European physicians practicing in Japan. The pathological research received little in the way of stimulus as only six autopsies could be undertaken in the years 1878 to 1880. In half of the dissected corpses, a disease other than beriberi was the cause of death, so only three postmortems could really contribute to beriberi research. Furthermore, all documented pathological studies were carried out at the university—not at the Beriberi Hospital itself—by a foreign physician, the German lecturer Erwin Baelz (1849–1913) who had been appointed Professor of Internal Medicine in 1876.[36]

While the research program based on hospital medicine did not succeed in identifying the cause of beriberi, the Beriberi Hospital nonetheless deeply influenced the Japanese medical community because it shaped a network of physicians committed to the conceptualization of beriberi as an infectious disease. As many of the doctors that had been affiliated with the Beriberi Hospital would later rise to positions of leadership in Japanese academia and the medical and military administrations, they would exert considerable influence in favor of research and health measures based on their causal perception. However, much to their irritation, it was the epidemiological approach based on nutritional theories that increasingly received attention during the following years.

Epidemiological Work and Experiments with the Barley–Rice–Diet

The naval surgeon Takagi Kanehiro (1849–1920) had been interested in beriberi since his youth. His father, who at the end of the Edo period had served in a military unit that guarded the Imperial palace in Kyoto, had told him about the disease that had cost many samurai their lives. In the troops, the samurai had thought that the disease was caused by the food that they were given and they had called the packages in which their rations were delivered "beriberi boxes."[37] In 1872, when Takagi entered the Japanese navy as a medical officer, he was immediately confronted with beriberi because the disease affected one third of all sick navy sailors, and it was clear that mainly beriberi patients occupied the two naval hospitals.[38] In 1880, after his return from five years of study at St. Thomas Hospital in England, Takagi devoted much energy, as head of the Tokyo naval hospital, to research on beriberi. He found great differences in the prevalence of the disease between the crews of different warships. In his search for the reason for these discrepancies, he first examined clothing and shelter of the sailors, but they turned out to be mostly uniform and he thus excluded them from the list of possible explanations. Only in the provisions did he detect significant variations among the crewmembers because sailors were given cash allowances for the free purchase of foodstuff. Takagi, therefore, concentrated his efforts on the improvement of the sailors' rations.[39]

Takagi was reportedly motivated by thinking "of the future of our [Japanese] empire, because, if such a [bad] state of health went on without discovering the cause and treatment of beriberi our navy would be of no use in time of need."[40] In 1882, two events made Takagi's work even more pressing. First, during the journeys that the Japanese fleet undertook during the Korea incident of July and August 1882, up to a third of the crew of the great flagships fell sick with beriberi and the combat readiness of the Japanese navy was seriously called into question. Second, the intense threat posed by beriberi was demonstrated by the occurrences during a trip of the training ship *Ryūjō*. The *Ryūjō* set sail in December 1882 with a crew of 376 sailors, headed for New Zealand, South America, and Hawaii. During the ten months until its return to Japan, 169 persons contracted beriberi, 25 of whom lost their lives.[41] After these events, Takagi was granted the opportunity to personally explain his ideas about fighting beriberi to the emperor. Based on successful trials on patients at the naval hospitals in early 1883, Takagi succeeded in reforming navy provisions first from white rice to a Western diet, later to a mixture of barley and rice.

Takagi attempted to demonstrate the effectiveness of the new provisions for the prevention of beriberi quasi-experimentally. He asked that the warship *Tsukuba* follow the same route as the *Ryūjō*, and he successfully requested that this time it be fitted out with new provisions that included meat and condensed milk. When the *Tsukuba* returned to Japan in November 1884, of its crew of 333 men, only 16 had contracted beriberi. The prevalence statistics for the navy as a whole, too, apparently confirmed the effectiveness of Takagi's forceful measures: While from 1878 to 1883, the incidence of beriberi among sailors had been as high as 41 percent, this figure dropped to 13 percent in 1884.[42] In March 1885, Takagi personally reported to the emperor his progress in fighting beriberi in the navy.

Outside of the navy, too, Takagi promoted his belief that the cause of beriberi was found in the diet based on white rice. Already in 1883, he presented his ideas for the first time before the "Great-Japan Private Society for Hygiene" (*Dai-Nihon shiritsu eiseikai*). In 1884, he published a table with instructions for the prevention of beriberi by means of correct nutrition that was distributed to all prefectures.[43] In order to lend scientific support to his empirical findings, Takagi reverted to the older theory that beriberi was caused by too high a proportion of carbon and too low a proportion of nitrogen in the diet.[44] In January 1885, on the zenith of his success after the completion of the *Tsukuba*-experiment, Takagi advertised his ideas again in a lecture "On the Prevention of the Beriberi Disease" before the Society for Hygiene.[45] Faced with Takagi's successes in the navy, even medical officers in the army began changing provisions. After Horiuchi Toshikuni (1844–1895) had successfully introduced a barley–rice–mixture at his division based in Ōsaka from 1884, most other army units had followed suit by 1890.[46]

In early 1885, the Meiji government began formal proceedings to bestow on Takagi an honor in recognition of his extraordinary achievements in fighting beriberi. On February 5, the Decoration Bureau (*Kunshō-kyoku*) in charge of conferring orders, asked the Ministry of Education for an evaluation of Takagi's scientific work, and the Ministry in turn requested that the university provide expertise. In its answer sent to the Ministry on March 26 for forwarding to the Decoration Bureau, however, the university faculty strongly denied Takagi's ideas. The academics argued that the cause of beriberi could not be reduced to dietary factors alone and that beriberi was a communicable, miasmatic disease. It would be highly unlikely that somebody through one or two experiments of only a few months duration could discover the cause of this disease and develop a method of prophylaxis. Finally, the faculty cast doubt on the specificity of Takagi's preventive measures against beriberi by observing that an improved diet contributes to the prevention of almost *every* disease. The group of experts who took a stance against Takagi's work included Harada Yutaka (?–1894), Ikeda Kensai, Ishiguro Tadanori, Hashimoto Tsunatsune, and Miyake Hiizu—all formerly affiliated with the government's Beriberi Hospital—as well as Ōsawa Kenji (1852–1927), Ōgata Masanori, Ise Jōgorō (1852–?), and the foreign lecturers van der Hayden and Julius Scriba (1848–1905).

In addition to their negative memorandum, individual university faculty opposed Takagi's ideas publicly. Particularly Ōsawa, who had become Japan's first professor of physiology after a period of postgraduate studies in Germany, expressed his concern that the spread of Takagi's views would lead to confusion in Japanese society. Ōsawa possibly considered it a danger that Takagi's beriberi theories might be understood as an endorsement of the dietetic theories of Kanpō medicine, such as those proposed by Tōda Chōan. This appeared even more concerning as the representatives of Western medicine had been fighting hard to discredit Chinese-style medicine.[47] Ōsawa not only doubted Takagi's hypothesis, but also in a second step, using physiological data, tried to prove scientifically that the barley-rice-mixture proposed by Takagi was not superior to a pure rice diet.[48] However, the known epidemiological facts pointed out by the British doctor William Edwin Anderson (1842–1900) already in 1878 would have been sufficient to prove wrong Takagi's assumption of a dietetic imbalance between carbon and nitrogen as the cause of beriberi.[49]

Ishiguro, too, published a new monograph about the beriberi disease, *Kakke-dan* (1885), in which he strongly criticized Takagi's ideas. While he conceded that diet played an important part in causing beriberi, he refused to accept Takagi's theory that the disease could be explained *solely* by factors of nutrition. He could not imagine that Takagi's therapy of providing patients with a mixture of barley and rice would be of any use in the struggle against the disease. Instead, he recommended as preventive measures an enhanced ventilation of troop barracks, a general improvement of foodstuff, and plenty of exercise.[50] Since Ishiguro was one of the highest-ranking medical officers of the army, his outright denial of the reforms implemented in the navy effectively slowed the learning process in the army.

Takagi's reforms, based on epidemiological studies and quasi-experimental support, received much attention, but his nutritional theories threatened the concept of beriberi as an infectious disease cherished by many influential members of Japan's medical establishment at the university and the army. By successfully opposing Takagi's decoration and scientifically undermining the rationale that he gave for his reforms, the members of this establishment prevented official recognition of Takagi's theories and managed to keep the race for the highly contested cause of beriberi open. Experimental proof of the infection theory, that is, the discovery of an actual beriberi germ, would be the ultimate weapon to restore the balance of power between the two parties.

BERIBERI, GERM THEORY, AND EARLY JAPANESE BACTERIOLOGY

In their efforts to raise doubts about Takagi's apparent successes, his opponents appeared entirely vindicated when only a few weeks later, in April 1885, Ōgata Masanori published his discovery of a germ causing beriberi. Ōgata's discovery temporarily confirmed the germ hypothesis of beriberi and thus proved the theory of beriberi being an infectious disease championed by Takagi's opponents. How was it possible that only a few years after the unsuccessful application of hospital medicine at the Beriberi Hospital, Ōgata could present a discovery based on the germ theory using the even more advanced techniques of laboratory medicine?

After the closure of the Beriberi Hospital in 1882, great hopes were pinned on bacteriological methods that had been so successfully applied to medical research in Europe, to fulfill the commitment of understanding the infectious cause of beriberi. Baelz who had participated in the pathological research at the Beriberi Hospital and who had profited most from it, first began implementing this agenda in Japan. In August 1882, he published an article in German "On Infectious Diseases Prevalent in Japan" that drew on "almost 6 years of experience at the heavily patronized inner clinic and policlinic of the university hospital in Tokyo that during this period were run under my direction." Baelz stressed that beriberi would be a "miasmatic infectious disease" and pointed to the startling "analogy with malaria." He "most decidedly" opposed Wernich's view that suggested a connection between *Kakke* and pernicious anemia. While bacteria had "not yet been identified in the blood" of beriberi patients, Baelz thought "it not unlikely that a parasite which until now has just escaped our research, will be found in there [i.e., in the blood] or the tissue."[51] Regarding bacteriological investigations of beriberi, Baelz wrote:

> From our present [scientific] position, the conception of Kakke as a miasmatic infectious disease almost brings about the duty to find the supposedly organized poison; most likely is the expectation that it is a body belonging to the group of fission fungi ["Spaltpilze"]. Based on this assumption, already for many years I have been trying to find such a body, be it in the blood, be it in the mainly affected organs, the nerves. Until now in vain. However, I do not give up the hope with the help of the recently so much perfected methods, especially Koch's staining procedure, still to reach the aim anyway and [I] will therefore continue the microscopic investigations further. Several times I believed to have found a specific Micrococcus, but since the finding was different in different preparations, I do not yet dare to view the same as the cause of Kakke.[52]

While Baelz began applying bacteriological methods to beriberi research, how prepared was Japan's medical community for the advent of bacteriology? Bacteriological topics had been introduced first by Japan's journals of medicine. In the *Tokyo Medical Journal* of 1878, an unidentified author reported on "Methods to exterminate Bacteria" (*Bakuteria o bokumetsu suru no hō*). The writer stated that microorganisms were a product of fermentation processes that could be observed under a microscope at a magnification of 800 times. The author claimed that for a physician it was most important to know how to fight bacteria, and then he discussed how different substances had proven to be of varying usefulness.[53]

Miyake Hiizu's book "General Theory of Pathology" (*byōri sōron*) that was published in 1879 contributed greatly to a more detailed knowledge of bacteriological facts in Japan. In drafting the manuscript for this book that went through several editions and was widely read, Miyake consulted Felix Victor Birch-Hirschfeld's *Textbook of Pathological Anatomy*[54] in addition to four other foreign works. Birch-Hirschfeld's text included the most recent findings of bacteriological research,[55] and based on this, Miyake gave a detailed overview of "schistomycetes" ("fission fungi") under the heading "plant parasites."[56]

Robert Koch's (1843–1910) discovery of the tuberculosis bacillus was transmitted to Japan only with a few months delay when Baelz explained Koch's work to the university students immediately before the summer vacation of 1882.[57] A written account of Koch's work reached the Japanese medical press the following year when Sakaki Junjirō (1859–1939), a physician who studied in Germany at the time, briefly communicated the experimental proceedings of Koch and his theory of the causation of tuberculosis.[58] However, the implications of Koch's discovery were not immediately grasped by the entire Japanese medical community and articles concerning bacteriological topics remained rather an exception in Japanese medical reporting.[59]

Therefore, foreign doctors were the ones who first applied bacteriological techniques to beriberi research. Van der Hayden of Kōbe was among the first physicians in Japan who—based on microscopic inspections of the blood of beriberi patients—claimed that "bacteria" or "micrococci" were the cause of the disease. In 1882, a Japanese medical journal briefly reported that van der Hayden had observed changes in the presence of bacteria in the blood of beriberi patients in correlation with the progress of the disease.[60] American missionary doctor Wallace Taylor (1835–1923) received attention even beyond Japan's shores with his finding of "spores" in the blood of beriberi patients.[61] In autumn 1884, he decided to investigate their link with the disease preparing cultures of the suspected germ and infecting laboratory animals with it. The infected creatures soon exhibited symptoms that according to Taylor were similar to those of human beriberi patients. He observed that the germ that he called "Beriberi Spirilum" was present in rice where cooking would not destroy it. This fit well with the folk wisdom of the Japanese people that the cause of beriberi was found in rice.[62]

Meanwhile, the interest in bacteriology grew considerably in Japan. In the introduction to its series "Overview of the Discovery of 'Bacteria' " (*'Bakuteria' hakkensetsu no shūshū*) of spring 1883, the *Tokyo Medical Journal* noted that during the past few years in the West, bacteriological theories had been increasingly discussed. It was expected that they would radically change the development of medicine. The paper pointed out that the example of tuberculosis had shown how the face of pathology was completely altered by the discovery of bacteria.[63] When one year later, in May 1884, the same journal published a series of articles on "Methods for the Observation of Bacteria" (*Bakuteria kensatsuhō*), the work and proceedings of Koch and Louis Pasteur were introduced in detail.

Only two years after Koch's discovery, the stage was already set for the first bacteriological debate among Japanese physicians. In 1884, the army surgeon Watanabe Kanae (1858–?) announced his discovery of a "Micrococcus Beriberi." Watanabe's investigation was driven by his conviction that the cause of beriberi should not be left to discovery by someone from the Western hemisphere because the disease was mainly prevalent in the Eastern hemisphere. Watanabe claimed to already having recognized in 1881 that the cause of the disease was a bacterium. He reported that in August 1882, when investigating the blood of patients, he had successfully identified the germ and that he had now reached the firm conclusion that this "parasite" (*parashiitsu*) and the beriberi disease were in an "inseparable relationship" (*aihanaru bekarazaru kankei*). The number of germs, it was argued, would correlate with the gravity of the illness. In the blood of patients with severe beriberi symptoms, there were more micrococci than in the blood of those who had only mild complaints. In addition, Watanabe had confirmed that the germ was not found in the blood of healthy persons and of patients suffering from a different disease. Watanabe apparently followed some of the causal criteria postulated by Koch as he reported that he had also tried to transmit the disease to animals, but that this work was still in progress.[64]

Hiroi Komaji and two other physicians thoroughly reviewed Watanabe's claims. The three critics imagined beriberi to take its course from a physical "predisposition" (*soin*), which when the "cause" (*gen'in*) was added, would lead to the outbreak of the disease in which food, clothing etc. would form promoting "circumstances" (*shoin*). They concluded that the question whether beriberi originated in the "conditions of everyday life" (*seikatsuhō*) or was caused by a "specific germ" (*toku'i dokuso*) could not yet be decided. Although they themselves were of the opinion that the cause of beriberi was a specific microorganism, they doubted Watanabe's discovery and regarded the true germ as still unidentified. They argued that already in 1871, Dutch and British physicians had discovered a "fungus" that was later recognized as having already been known and that it was probably similar to Watanabe's discovery. There were many conditions that a proposed beriberi germ had to fulfill. Hiroi and his colleagues called for Watanabe to try the method that Koch valued so highly: to isolate the organism and then to infect laboratory animals with it. They concluded that at the present stage of research, foodstuff and clothing could still not be excluded as causes of the disease.[65]

ŌGATA'S DISCOVERY OF BERIBERI GERMS AND THE POWER OF "LABORATORY MEDICINE"

When Ōgata Masanori returned from postgraduate work in Germany in December 1884, Japanese physicians had already joined Baelz and other foreign colleagues in the hunt for the supposed beriberi germ. Japan's medical community was also sufficiently informed to critically evaluate hasty bacteriological discoveries. When upon his return, the university and the Ministry of the Interior both immediately employed Ōgata to head their respective bacteriological laboratories, he made it his highest priority to identify the cause of beriberi with the bacteriological techniques that he had studied in Germany.

Ōgata was the son of a family of physicians from Kumamoto in Kyushu where he began studying medicine before moving to the university in Tokyo. After graduation in 1880, he assisted Baelz with his work on beriberi. In January 1881, Ōgata received a government scholarship for study in Germany where he first concentrated on physiology and hygiene at the University of Leipzig, Baelz' Alma Mater. Later, Ōgata moved to Munich where he continued research on hygiene with Pettenkofer.[66] In 1884, two years after Koch's discovery of the tuberculosis bacillus, Ōgata spent several months in Berlin to learn bacteriological techniques at the Reichsgesundheitsamt. It is likely that Ōgata acted on orders from Nagayo, the powerful chief of the Bureau of Hygiene, because the Ministry of the Interior offered to bear his expenses during his stay in Berlin.[67] Since Koch was visiting Egypt and India at that time, his assistant, Friedrich Löffler (1852–1915), initially instructed Ōgata.[68]

Ōgata's laboratory in Tokyo had already been partially prepared with government help upon his return: Shibata Tsuguyoshi (1850–1910) of the

Ministry of the Interior who had visited Berlin to attend the hygiene exhibition of 1883, had transported part of the valuable equipment on his return trip to Japan; Ōgata brought the rest with him.[69] He received blood and tissue samples from the beriberi department at the university[70] and examined them at the Tokyo Laboratory for Hygiene of the Ministry of the Interior that was equipped with three Zeiss microscopes with immersion lenses for maximum magnification.[71] At the university, Tsuboi Jirō (1862–1903) was assisting Ōgata while at the Tōkyō Laboratory, he was aided by Kitasato who had concluded his medical studies at the university three years after Ōgata. In addition, Ōgata was expected to train three physicians—Kako Tsurudo (1855–1931) of the army, Kuwahara Sōsuke of the navy, and Suga Yukiyoshi (1854–1914) of the Okayama Medical School—in bacteriological techniques.[72]

Only four months after his return from Germany and only a few weeks after Takagi's report on his nutritional experiments, the great investments in Ōgata's education and research appeared to pay off when on April 6, 1885, Ōgata published in the official government gazette (*kanpō*) a formal "Report about the Discovery of the Beriberi Bacillus" (*Kakke byōkin hakken no gi kaishin*).[73] Ōgata claimed to have isolated a hitherto unknown microorganism from the blood of beriberi patients and the tissue of deceased beriberi victims. Ōgata declared that he could breed this microbe in pure culture and that after inoculation in laboratory animals, it produced symptoms and pathological signs that closely resembled those of beriberi patients.[74] In composing his report, Ōgata emphasized from the beginning that he had followed Koch's example by successfully isolating the bacillus *and* infecting laboratory animals with it and that he had therefore concluded that it was the cause of disease.[75] In a short span of time, Ōgata had thus successfully raised the quest for the cause of beriberi to a new level by bringing the pinnacle of scientific medicine, laboratory medicine, to bear on this task.

During the following weeks, Ōgata held two public lectures about his discovery at the invitation of the president of the university, Katō Hiroyuki (1836–1916), and the director of the Bureau of Hygiene, Nagayo. Among the audience were not only faculty members of the university, but also leading representatives from government, medicine, and the military.[76] In front of a blackboard with explanatory drawings, Ōgata had installed microscopes through which the visitors could observe his "beriberi bacillus." In addition, cultures of the bacilli growing on different media were exhibited. In his speech, Ōgata explained his methods of investigation in detail. To further substantiate his findings, he also presented laboratory animals whose hind extremities were paralyzed, apparently in a way characteristic of the symptoms of beriberi.[77] In quickly presenting his preliminary results to the public, Ōgata thus made intensive use of many of the new forms of visual and "functional" representation of his "discovery" that communicated laboratory medicine's claim to objectivity.[78]

At the end of his presentation, Ōgata turned to his competitors. After criticizing aspects of Taylor's work, he particularly stressed the fundamental differences between the implications of his discovery and the theories of the also present Takagi.[79] After Ōgata had finished, Takagi had the opportunity to respond. In the face of Ōgata's overwhelming experimental evidence, Takagi attacked from a pragmatic viewpoint: Ōgata's discovery was not very practical, because if it held true, then all physicians would have to be equipped with expensive microscopes to diagnose beriberi with certainty. In addition, Takagi doubted that Ōgata's research would lead to an improvement of beriberi treatment. This argument was indeed powerful. Already in 1881, an essay about the beriberi disease by Baelz had disappointed the Japanese readership because the author did not derive recommendations for therapy from his bacteriological theories.[80] The publishing house resorted to printing advice from an unidentified source in the next edition of the journal.[81] As Takagi did not have the training needed to directly question Ōgata's laboratory evidence, he chose to contest Ōgata's results on the grounds of usefulness instead.

Finally, Ishiguro addressed the audience and lavishly praised Ōgata's discovery. According to Ishiguro, Ōgata had used such precise research methods as had been unknown to "oriental people" (*tōyōjin*) and most of the Western physicians practicing in East Asia. Ishiguro was also deeply impressed by the opposing views of Ōgata and Takagi, both of whom were his personal friends. His speech ended with an appeasing gesture stressing the stimulating effect that differences in opinion would have on true scientists.[82]

Ōgata's discovery left a deep impression on the medical community in Japan, even causing a small bacteria boom. Already a few weeks later, the *Tokyo Medical Journal* reported that Joseph Disse (1852–1912), a German lecturer at the university, had also discovered a beriberi "fungus" that resided at different locations in the spinal chord.[83] In a letter to the journal, Taylor once more called attention to his discovery of the "Beriberi Spirilum."[84] Ōgata himself continued his investigation of the "beriberi bacillus" that he also proudly presented in a German medical weekly in the same year.[85] After having been appointed professor at the university to teach hygiene,[86] one year later, he published a second report about his work on the beriberi germ.[87]

In spite of the generally favorable response to Ōgata's discovery, many Japanese doctors still harbored reservations and judged his findings not yet sufficiently confirmed. In response to a question concerning the beriberi disease, Yamazaki Motomichi of the Society for Hygiene for example, answered that the cause and pathology of beriberi were still unknown. He himself believed that Ōgata's bacillus was indeed the cause of the disease, but that this result still awaited validation. Moreover, Yamazaki combined the germ theory with the older miasmatic disease concept explaining that beriberi was an infectious disease that was contracted from the soil.[88] In a monograph on beriberi published by Harada, the author also stuck to the hypothesis of a miasmatic infectious disease.[89] Ōgata's discovery was thus smoothly integrated into a germ concept that differed from Koch's: Many Japanese physicians believed germs to be miasmatic in origin. Like many of their European colleagues, most Japanese doctors embraced Koch's concept of specificity—that pathogenic organisms could be produced only from organisms of the same species—only many years later.[90]

While Japan's medical circles were not uncritical, it was hard not to be impressed when confronted with Ōgata's cutting-edge laboratory methods that were modeled on Koch's example, in combination with the authority that Ōgata's study at the Reichsgesundheitsamt in Berlin had conferred on him. Physicians based in Japan did not dare to challenge Ōgata's findings as forcefully as they had done before with Watanabe's claims. This role finally fell to Kitasato who for a short period had been Ōgata's assistant in preparing the discovery of the "beriberi bacillus" before leaving to work with Koch in Berlin. There, Kitasato had the chance to study the methods of the new laboratory science over a much longer period than Ōgata had done. This put Kitasato in a position to criticize Ōgata's work on beriberi as a specialist of bacteriology, and this lead in 1888 to the much-discussed controversy between him and Ōgata. However, being a bacteriologist himself, Kitasato did not explicitly doubt that the cause of beriberi was a germ; he only questioned that *Ōgata*'s "beriberi bacillus" was that germ.

Supported by laboratory medicine, Ōgata's discovery gave the physicians championing the infectious disease theory of beriberi nonetheless more than just a short-lived opportunity to draw attention away from Takagi's practical successes. The doctors at the university and in the army found lasting support for their position through Ōgata's discovery because by introducing laboratory methods, Ōgata had raised the demands placed on a scientifically acceptable causal explanation to a level that Takagi could not match. While the nutritional origin of beriberi postulated by Takagi was too unspecific to satisfy the standards of evidence inspired by "classical bacteriology," Takagi's work did not yet exhibit the combination of work in epidemiology and hygiene with the laboratory search for a specific cause that became characteristic of tropical medicine after 1900.[91] Proposing a cure without being able to establish a suitable cause put Takagi in a position similar to that of the ousted Kanpō physicians with their time-tested therapies based on speculative theories. In the navy, the incidence of beriberi continued to drop rapidly from two-digit levels to 0.6 percent in 1885 and even 0.1 percent in 1886, and Takagi's practical successes received international recognition,[92] but he could win only a few followers because of his unconvincing theoretical explanation. By raising the standards of what is scientific, Ōgata's discovery had effectively shifted the balance in the debate.

CONCLUSIONS

The integration of medicine into Japan's modernization policies from the middle of the nineteenth century[93] found its expression in the particular arena of beriberi in the intensive search for the cause of this disease that prompted the government to step in and seek to control it. However, after a government-sponsored research program of hospital medicine failed to identify a specific cause of beriberi, the "true" origin of the disease remained contested between physicians favoring empirical conceptions based on nutrition and those believing in theories of infection as proposed by many representatives of Western medicine. In the race for the identification of the cause of beriberi that ensued, both sides, using government resources, turned to experimental approaches to prove their ideas. The physicians, preferring the evolving germ theory of beriberi, countered the practical successes of their competitors with findings produced with the modern methods of scientific medicine that they speedily introduced into Japan. In the field of beriberi research, Japan completed the transition from hospital medicine to laboratory medicine, which had taken many decades in the West, in only seven years. Although the discoveries made in the laboratory were met with skepticism, the outcome of this contest was a lasting stalemate in which supporters of nutritional concepts succeeded in implementing prevention measures against beriberi while backers of germ theory managed to block official recognition of nutritional ideas.

The adoption of scientific medicine with its experimental approaches in Japan was strongly driven by the perceived economic and military need to control endemic beriberi, facilitating the supply of massive state resources. Individual physicians from both conceptual camps involved in beriberi research repeatedly stated that they regarded their work as being of national importance. However, perceiving the control of beriberi as a precondition for military preparedness was not peculiar to the Japanese. During the modernization process in many countries, efforts to fight disease created a rising interest in the identification of necessary causes whose removal could serve as preventive measures.[94] The Dutch colonial authorities in Indonesia, for example, saw "conquering beri-beri [. . .] as a necessary condition for winning the Atjeh wars" and this perception formed the background for the mission by Pekelharing and Winkler arriving in Java in 1886 to investigate the disease's cause.[95] In Japan, the physicians at the state's Beriberi Hospital also sought to find a cause whose removal would allow the control of the disease—a necessary cause. Thus in Japan at the end of the 1870s and in the early 1880s, a similar shift from primarily considering disease symptoms to a concern with etiology can be observed, as it has been pointed out for beriberi research published in the Dutch language around the 1880s.[96] This interest in the *cause* of beriberi provided the link between the perceived need to control the disease and the ensuing research agenda that led to the ready adoption first of hospital medicine and then of laboratory medicine in Japan.

While hospital medicine soon reached its limits because the desired findings were not readily forthcoming, it prepared the ground for the next stage of the quest into beriberi's causation by forging an influential group of physicians supporting the model of infection. When the temporary void left by the lack of practical results from the Beriberi Hospital was filled by the nutritional approach advanced through experimental means, the leap from hospital to laboratory medicine was quickly taken. Germ theory arrived on the scene at a moment when Japan's medical elite was acutely absorbed in the search for the origin of beriberi. However, the transition from hospital to laboratory medicine was possible not just because germ theory had penetrated the Japanese medical community during the preceding years, but also because at this moment the Japanese government's program to send students abroad produced a person—Ōgata Masanori—who seemed fully equipped to successfully implement the most advanced program of research available at the time, laboratory medicine.

The acceptance of a beriberi germ by the Japanese medical community was helped by its apparent close resemblance to indigenous theories of a "wind poison" evaporating from the soil and entering the body via the feet, that appeared compatible with the disease's particular epidemiology, and that was supported by influential foreign physicians believing beriberi to be a miasmatic infection. The rapid adoption of the germ theory in Japan in the context of beriberi research is thus also an instructive example of local appropriation and demonstrates that "there was no 'germ theory of disease' transcendent over time, but rather many different germ theories of specific diseases being debated in specific communities, times, and places [, . . . and] particular understandings of the germ theory were [indebted] to preexisting traditions of explaining disease."[97] It also exemplifies the difficulty of the diffusion of a highly codified scientific discipline, even under the conditions of the seemingly well organized Japanese modernization process. Even in Japan, with its hired foreign teachers and its great number of physicians studying at leading academic institutions abroad, the "laboratory practice that developed [. . .] in the first wave of enthusiasm for the 'miracle-making' science [bacteriology] often failed to conform to the discipline's new, more stringent professional standards."[98] Therefore during the early development of bacteriology in Japan, debates centered on the technical aspects of bacteriological work, and like similar discoveries of beriberi germs for example in South America,[99] early Japanese announcements were rejected because they did not conform to the high standards of bacteriological research. However, these technical "teething" problems were largely overcome after researchers like Kitasato returned to Japan who had the opportunity to undergo much more in-depth training in the new scientific methods than their predecessors.

While the will to remove beriberi as an obstacle to Japan's modernization and the ensuing struggle over the disease's causal explanation accelerated the introduction of scientific medicine to Japan, this did not necessarily bring the "fruits of progress" to the Japanese people. Especially in the army's medical corps, physicians committed to the germ theory of beriberi and supported by bacteriological findings continued to exert a strong influence. Ishiguro in particular repeatedly resisted attempts to reform the army's rice-based diet, ultimately at great cost. Ten years after Ōgata's discovery, during the Sino-Japanese war of 1894–1895, casualties caused by beriberi were nine times higher than those due to combat action. And in 1904–1905, when Japan's victory over a major European power in the Russo-Japanese war was celebrated by many Japanese as proof of the success of Japan's modernization policy, this triumph was tarnished by the fact that more than

200,000 Japanese army soldiers or almost 20 percent of total army personnel in the field in Japan and Asia fell sick with beriberi, many of them dying from the disease.[100] Several decades after the Meiji Restoration and the beginning of Japan's modernization policy, modern medicine provided leading physicians with a scientific rationale to effectively oppose prevention measures against beriberi, the effectiveness of which had been demonstrated "only" empirically.

NOTES

1. Christian Oberländer, *Zwischen Tradition und Moderne: Die Bewegung für den Fortbestand der Kanpō-Medizin in Japan* (Stuttgart: Franz Steiner [Medizin, Gesellschaft und Geschichte, Beiheft 7], 1995), pp. 51–65.
2. Andrew Cunningham and Bridie Andrews, "Introduction: Western Medicine as Contested Knowledge," in *Western Medicine as Contested Knowledge*, ed. Andrew Cunningham and Bridie Andrews (Manchester: Manchester University Press, 1997), pp. 1–23; here pp. 8–9, 12.
3. Andrew Cunningham and Perry Williams, "Introduction," in *The Laboratory Revolution in Medicine*, ed. Andrew Cunningham and Perry Williams (Cambridge: Cambridge University Press, 1992), pp. 1–13; here pp. 2–5.
4. Personal names are given in the customary order in the native language of the person. Where they are known, the years of birth and death of people are given.
5. See, e.g., Yamashita Seizō, *Meijiki ni okeru kakke no rekishi* (History of the Beriberi Disease in the Meiji Period) (Tokyo: Tōkyō Daigaku Shuppankai, 1988), p. 295; Itakura Kiyonobu, *Mohō no jidai* (The Age of Imitation) (Tokyo: Kasetsusha, 1988), p. 299.
6. Fujino Tsunezaburō, *Fujino, Nihon saikingaku-shi* (Fujino's History of Japanese Bacteriology) (Tokyo: Kindai Shuppan, 1984), pp. 105–114.
7. James Bartholomew, *The Formation of Science in Japan* (New Haven: Yale University Press, 1989), p. 81.
8. K. Cordell Carter, "The Germ Theory, Beriberi, and the Deficiency Theory of Disease," *Medical History* 1977, *21*: 119–136.
9. Oberländer, *Zwischen Tradition und Moderne*, pp. 61–64.
10. Yamashita Seizō, *Kakke no rekishi: bitamin hakken izen* (History of Beriberi: Before the Discovery of the Vitamin) (Tokyo: Tōkyō Daigaku Shuppankai, 1983), pp. 183, 191, 220, 356–358.
11. Heinrich Botho Scheube, "Die japanische Kak-ke (Beri-beri)," *Deutsches Archiv für klinische Medizin* 1882, *31*, 1 and 2 (May 30): 141–202; 3 and 4 (July 13): 307–348; *32*, 1 and 2 (November 8): 83–119; here pp. 148–149.
12. Yamashita, *Meijiki ni okeru kakke no rekishi*, pp. 89, 335–336.
13. Ibid., pp. 24–27, 43.
14. Kōseishō Imukyoku, *Isei hyakunenshi shiryōhen* (Hundred Year History of the Medical Law: Sources) (Tokyo: Gyōsei, 1976), pp. 52–53.
15. Oberländer, *Zwischen Tradition und Moderne*, pp. 86–92.
16. Yamashita, *Meijiki ni okeru kakke no rekishi*, pp. 260–261.
17. William Johnston, *The Modern Epidemic: A History of Tuberculosis in Japan* (Cambridge, Mass.: Council on East Asian Studies, Harvard University, 1995), p. 188.

18. Ibid., p. 188.
19. Kashimura Seitoku, "Kakke-ron" (On Beriberi), *Tōkyō iji shinshi* (Tokyo Medical Journal) April 10, 1878, *16*: 5–13; here pp. 5, 10.
20. Hashimoto Tsunatsune, "Kakke shinsetsu" (New Theory of Beriberi), *Iji shinbun* May 11, 1878, *1*: 1–13; here pp. 2–4.
21. In addition to the term "bacteria," other expressions were frequently used in Japan. For example, the term "fungus" (*pirutsu*) was common. Hashimoto Tsunatsune described the pathogen of diphteria as a "fungus" that enters the mouth from the atmosphere (Hashimoto Tsunatsune, "Kōtō 'Jifuterichisu' no setsu" [On "Diphteria" of the Throat], *Tōkyō iji shinshi* February 22, 1879, *48*: 1–13; here pp. 4–12).
22. Ishiguro Tadanori, *Kakke-ron* (Theory of Beriberi) (Tokyo: Eirandō, 1878), pp. 3, 5, 21.
23. For a more complete overview of the theories of beriberi's causation that Japanese and foreign physicians of the Meiji period proposed, see Christian Oberländer, "The Rise of Scientific Medicine in Japan," *Historia Scientiarum* 2004, *13* (3): 176–199; here pp. 177–180.
24. Oberländer, *Zwischen Tradition und Moderne*, pp. 83–84; Yamashita, *Meijiki ni okeru kakke no rekishi*, pp. 95–97, 100.
25 Kakke Byōin, *Kakke byōin daiichi hōkoku* (First Report of the Beriberi Hospital) (Tokyo: Kakke Byōin, 1879).
26. Ibid.
27. Heinrich Botho Scheube, "Die japanische Kak-ke (Beri-beri)," *Deutsches Archiv für klinische Medizin* 1882, *31*, 1 and 2 (May 30): 141–202; 3 and 4 (July 13): 307–348; *32*, 1 and 2 (November 8): 83–119; here p. 147.
28 Yamashita, *Meijiki ni okeru kakke no rekishi*, pp. 179–181, 185, 193, 196, 207.
29. Kakke Byōin, *Kakke byōin daiichi hōkoku*, pp. 89–90, 92.
30. Ibid., pp. 90–92.
31. "Naika senmon shokai" (Internistic Meeting), *Iji shinbun* (Medical Newspaper) August 15, 1880, *29*: 1–21.
32. Satomi Giichirō, "Miasma shobyō" (Miasmatic Diseases), *Iji shinbun* September 15, 1880, *30*: 1–3.
33. Kakke Byōin, *Kakke byōin daini hōkoku* (Second Report of the Beriberi Hospital) (Tokyo: Kakke Byōin, 1881), p. 77.
34. Ibid., pp. 117–118.
35. Yamashita, *Meijiki ni okeru kakke no rekishi*, p. 229 note 22.
36. Heinrich Vianden, *Die Einführung der deutschen Medizin im Japan der Meiji-Zeit* (Düsseldorf: Triltsch Verlag [= Düsseldorfer Arbeiten zur Geschichte der Medizin 59], 1985), p. 134.
37. Takaki Kanehiro, "Three Lectures on the Preservation of Health amongst the Personnel of the Japanese Navy and Army. Lecture I," *The Lancet* May 19, 1906: 1369–1374; May 26: 1451–1455; June 2: 1520–1523; here p. 1370.
38. Yamashita, *Meijiki ni okeru kakke no rekishi*, p. 334.
39. Ibid., pp. 339–340.
40. Takaki, "Three Lectures," p. 1370.
41. Yamashita, *Meijiki ni okeru kakke no rekishi*, p. 338.
42. Ibid., pp. 333, 343–352.
43. "Kakke gen'in" (Cause of the Beriberi Disease), *Tōkyō iji shinshi* December 27, 1884, *352*: 1666.

44. Yamashita, *Meijiki ni okeru kakke no rekishi*, pp. 339–340.
45. Takagi Kanehiro, "Kakke-byō yobō-setsu" (About the Prevention of the Beriberi Disease), *Dai-Nihon shiritsu eiseikai zasshi* (Journal of the Great-Japan Private Society for Hygiene) 1885, *22*: 1–20.
46. Yamashita, *Meijiki ni okeru kakke no rekishi*, pp. 399–401.
47. Oberländer, *Zwischen Tradition und Moderne*, pp. 65–106.
48. Ōsawa Kenji, "Bakuhan no setsu" (About the Barley-Rice Mix), *Dai-Nihon shiritsu eiseikai zasshi*, July 18, 1885, *26*: 1–13 and August 18, *27*: 1–16.
49. William Anderson, "Kak'ké," *Transactions of the Asiatic Society of Japan* October 27, 1878, *6* (1): 155–178; here pp. 155, 169–170, 175. An overview is given by Carter, "The Germ Theory," pp. 126–127.
50. Ishiguro Tadanori, *Kakke-dan* (About Beriberi) (Tokyo: Eirandō, 1885).
51. Erwin von Baelz, "Ueber die in Japan vorkommenden Infectionskrankheiten," *Mittheilungen der OAG* August 1882, *27*: 295–319; here pp. 304–307, 315.
52. Ibid., p. 304.
53. "Bakuteria o bokumetsu suru no hō" (Methods to exterminate Bacteria), *Tōkyō iji shinshi* January 25, 1878, *12*: 14–17.
54. Felix Victor Birch-Hirschfeld, *Lehrbuch der pathologischen Anatomie* (Leipzig: F.C.W. Vogel, 1877).
55. K. Cordell Carter, "Koch's Postulates in Relation to the Work of Jacob Henle and Edwin Klebs," *Medical History* 1985, *29*: 353–374; here p. 365.
56. Fujino, *Fujino, Nihon saikingaku-shi*, pp. 44–45.
57. Ibid., p. 91.
58. Sakaki Junjirō, "Kekkakusho ha hatashite densenbyō nari" (Is Tuberculosis really an Infectious Disease?), *Tōkyō iji shinshi* May 5, 1883, *266*: 12–16; here p. 13.
59. Johnston, *The Modern Epidemic*, p. 191.
60. "Kakke kanja no ketsueki kensa" (Examination of the Blood of Beriberi Patients), *Tōkyō iji shinshi* July 8, 1882, *223*: 30–31.
61. See, e.g., "Kakké, or Japanese Beri-beri," *Lancet* June 30, 1887: 233–234; here p. 234.
62. Wallace Taylor, "Kakke ichimei beri-beri no gen'in" (The Cause of *Kakke* or Beriberi), *Tōkyō iji shinshi* August 8, 1885, *384*: 998–1001.
63. " 'Bakuteria' hakkensetsu no shūshū" (Overview of the Discovery of "Bacteria"), *Tōkyō iji shinshi* April 28, 1883, *265*: 5–10; May 12, *267*: 8–12; May 19, *268*: 6–9; here p. 5.
64. Watanabe Kanae, "Kakke byōdoku hatsumei-ron" (About the Discovery of the Beriberi Agent), *Tōkyō iji shinshi* September 27, 1884, *339*: 1207–1211 and October 4, *340*: 1241–1247; here pp. 1208–1211; 1242–1246.
65. Hiroi Komaji et al., "Kakke gen'in-ron" (About the Cause of Beriberi), *Chūgai iji shinpō* (International Medical Review) January 10, 1885, *115*: 15–19; January 25, *116*: 23–25; February 10, *117*: 20–22; February 25, *118*: 19–22.
66. Itakura, *Mohō no jidai*, p. 290.
67. Fujino, *Fujino, Nihon saikingaku-shi*, p. 102.
68. Itakura, *Mohō no jidai*, p. 290.
69. Fujino, *Fujino, Nihon saikingaku-shi*, pp. 102–103.
70. Ōgata Masanori, "Kakke byōdoku hakken" (Discovery of the Beriberi Disease Poison), *Tōkyō iji shinshi* April 11, 1885, *367*: 454–457; April 18, *368*: 492–497; April 25, *369*: 517–522; here p. 454.
71. Fujino, *Fujino, Nihon saikingaku-shi*, p. 105.
72. Ibid., p. 104.

73. Yamashita, *Meijiki ni okeru kakke no rekishi*, p. 298.
74. Ōgata, "Kakke byōdoku hakken," pp. 454–455.
75. Carter, "Koch's Postulates," p. 361.
76. "Kakke byōdoku hatsumei dai-enzetsukai kiji" (Report on the Great Lecture Event concerning the Discovery of the Beriberi Disease Poison), *Tōkyō iji shinshi* April 18, 1885, *368*: 507–510.
77. Ibid., pp. 507–508.
78. For the use of different methods of representation in bacteriology, see e.g., Thomas Schlich, "Linking Cause and Disease in the Laboratory: Robert Koch's Method of Superimposing Visual and 'Functional' Representations of Bacteria," *History and Philosophy of the Life Sciences* 2000, *22*: 43–58.
79. "Kakke byōkin hakken enzetsu" (Lecture on the Discovery of the Cause of the Beriberi Disease), *Chūgai iji shinpō* April 25, 1885, *122*: 24–26; here p. 25; "Dainikai kakke baikin enzetsu" (Second Lecture on the Beriberi Germ), *Chūgai iji shinpō* May 10, 1885, *123*: 25–27; here p. 26.
80. Erwin von Baelz, "Kakkebyō-ron" (About the Beriberi Disease), *Chūgai iji shinpō* March 25, 1881, *26*: 1–8 and April 10, *27*: 1–10.
81. "Igaku shinsetsu" (New Medical Theories), *Chūgai iji shinpō* April 25, 1881, *28*: 1–3. Concerning the discovery of the microorganism causing tuberculosis, too, some authors criticized that this would not contribute to the treatment of the disease (Johnston, *The Modern Epidemic*, p. 191).
82. "Kakke byōdoku hatsumei," pp. 509–510.
83. "Isshū no kin" (A Kind of Bacteria), *Tōkyō iji shinshi* May 2, 1885, *370*: 578.
84. Taylor, "Kakke ichimei beri-beri no gen' in," p. 998.
85. "Untersuchungen über die Aetiologie der Kakke," *Aerztliches Intelligenzblatt* November 24, 1885, *32*, *47*: 683–686.
86. Yamashita, *Meijiki ni okeru kakke no rekishi*, p. 298.
87. "Kakke byōgen kensa" (Investigation of the Cause of Beriberi), *Tōkyō iji shinshi*, April 3, 1886, *418*: 428–433; April 10, *419*: 465–470; April 17, *420*: 501–505; April 24, *421*: 537–544; May 1, *422*: 571–576; May 8, *423*: 600–607; May 15, *424*: 634–641.
88. Yamazaki Motomichi, "Kakkebyō ōtō" (Answers concerning the Beriberi Disease), *Dai-Nihon shiritsu eiseikai zasshi* July 25, 1885, *26*: 53–58; here p. 53.
89. Itakura, *Mohō no jidai*, p. 250.
90. For details concerning the debate about specificity during the early development of bacteriology, see Pauline Mazumdar, *Species and Specificity. An Interpretation of the History of Immunology* (Cambridge: Cambridge University Press, 1995).
91. Ilana Löwy, "Yellow Fever in Rio de Janeiro and the Pasteur Institute Mission (1901–1905): The Transfer of Science to the Periphery," *Medical History* 1990, *34*: 144–163; here p. 162.
92. See, e.g., Takaki, "Three Lectures."
93. Oberländer, *Zwischen Tradition und Moderne*, pp. 44–45.
94. K. Cordell Carter, "The Development of Pasteur's Concept of Disease Causation and the Emergence of Specific Causes in Nineteenth-Century Medicine," *Bulletin for the History of Medicine* 1991, *65*: 528–548; here p. 544. See also Thomas Schlich, "Die Konstruktion der notwendigen Krankheitsursache: Wie die Medizin Krankheit beherrschen will," in *Anatomien medizinischen Wissens. Medizin, Macht, Moleküle*, ed. Cornelius Borck (Fischer: Frankfurt a.M., 1996), pp. 201–229.

95. Harmke Kamminga, "Credit and Resistance: Eijkman and the Transformation of Beri-beri into a Vitamin Deficiency Disease," pp. 232–254; here p. 236.
96. Ibid., p. 238.
97. Nancy J. Tomes and John Harley Warner, "Introduction to the Special Issue on Rethinking the Reception of the Germ Theory of Disease: Comparative Perspectives," *Journal of the History of Medicine* 1997, *52*: 7–16.
98. Löwy, "Yellow Fever," p. 144.
99. Ibid., p. 144.
100. Yamashita, *Meijiki ni okeru kakke no rekishi*, pp. 440–465.

2

Male Anxieties: Nerve Force, Nation, and the Power of Sexual Knowledge

*Sabine Frühstück**

Introduction

In the fall of 1929, a Kyōto-based journal for popular medicine reported that the dean of sexology, Habuto Eiji, had committed suicide after having long suffered from neurasthenia (*shinkei suijaku*).[1] A practicing gynecologist, Habuto had been the editor of the sexological journal *Seiyoku to Jinsei (Sexual Desire and Humankind)*, the author of numerous books on sexual issues, and the coauthor, together with Sawada Junjirō, of an abridged Japanese version of Richard von Krafft-Ebing's *Psychopathia Sexualis*, entitled *Hentai Seiyokuron* (1915). He was also involved in the translation of Havelock Ellis's *Studies in the Psychology of Sex* (1901–1928), the twenty Japanese-language volumes of which were advertised under the title *Sei no Shinri* (see figure 2.1) as early as 1927.

Among other sexologists, Habuto had been a chief theorist on the causes of neurasthenia. Physicians, psychiatrists, psychologists, pedagogues, and sexologists agreed with him that neurasthenia primarily afflicted men and was caused by overpowering exhaustion that was, in turn, the result of certain sexual practices. While early Japanese treatises on similar phenomena had attributed them primarily to abstinence or to "the lack of unification between man and woman,"[2] modern commentators like Habuto speculated that neurasthenia was the result of masturbation or—even worse—homosexuality.[3]

For the journal to report neurasthenia as the cause of Habuto's death seemed ironic, then, considering that Habuto Eiji had thundered against both masturbation and homosexuality for most of his life. In the spirit of his time, Habuto had believed in the presupposed connection between the health of the individual body and the security of the nation. Habuto and other sexologists were certain that manhood and national security could be achieved only by educating boys and men about the dangers of a variety of sexual behavior. He was certain that neurasthenia could have disastrous effects not only on the health of the affected individual but also on the

Figure 2.1 This pamphlet advertises a 20-volume, abridged Japanese version of Havelock Ellis's *Studies in the Psychology of Sex*, which he had begun to publish in 1900 as *Sei no shinri* (1927).

welfare and strength of the Japanese body politic. To scientific knowledge, whether sexual, medical, psychological, or any other kind, he attributed a certain power. In Habuto's and in many of his contemporaries' minds, the exercise of political power was to be informed by scientific knowledge. Consequently, he insisted that the physical, mental, and political empowerment of the ignorant masses was possible only through education.

Habuto was an important figure in the creation of the complicated texture of medical, psychological, and pedagogical theories on human sexuality in early-twentieth-century Japan, which I attempt to untangle in this chapter. I argue that the emerging science of sex (*seikagaku* or *seigaku*) simultaneously contributed to and shaped a new understanding of manhood and of the formation of the modern Japanese nation. These theories included a medical understanding of "nerve force" as a major component of mental health; a psychology that primarily dealt with pathologies of the will, including manias, hysteria, and neurasthenia as well as questionable and contradictory healing methods; an understanding of disease as intrinsically tied to modernity and progress; and utopias of masculinity that constructed the ideal male body as resembling the nation in terms of its mental/political strength and its sexual/military potency as well as its countertype, the mentally deranged and physically weak man.

During the late nineteenth century, neurasthenia emerged as a new amorphous "disease entity" which metamorphosed several times during the following decades.[4] Constructed within the boundaries of military medicine as a catchall category for minor mental dysfunctions in the 1880s, neurasthenia was reframed by experts in pedagogics and psychiatry at the beginning of the twentieth century. During the 1920s and 1930s, sexologists like Habuto redressed the set of ailments that had been associated with neurasthenia as a problem of sexual behavior that threatened men's health and—by implication—Japan's social order and national stability. The popular medicine of the late 1940s and 1950s then cleansed neurasthenia from its sexual and pathological touches and pushed for an understanding of overwork as its exclusive cause.

Neurasthenia underwent an impressive career across scientific disciplines as well as through social and political realms. Understood by imperial army surgeons as a male phenomenon that impeded military performance, the causes of neurasthenia were believed to include masturbation, sexual immorality as well as sexual abstinence, and overwork. Its symptoms ranged from paleness, loss of appetite and forgetfulness to melancholy, low work efficiency and a general weakening of body and mind.[5] Its effects included homosexuality, syphilis, tuberculosis, and suicide. Thus, ignorance of neurasthenia was considered dangerous, both to individual masculinity and to the defense against Western colonial powers in particular, and the challenges of modernity in general.[6] Pedagogues, by contrast, thought neurasthenia to be common among boys and girls, even though it remained more worrisome when occurring in boys. Medical handbooks for home use, published in the 1950s, again declared overwork as a cause of neurasthenia, and that it was a male problem. Illustrations in these encyclopedias that depicted men working at their desks late at night also suggested that neurasthenia had become a male, white-collar phenomenon that apparently did not bother women or the working classes.

Challenged on a number of fronts, the concept of masculinity that emerged from the history of neurasthenia between the 1870s and the mid-twentieth century demanded constant work.[7] In Japan, I argue, the nationalism around the turn of the nineteenth century provided a powerful base for a manly ideal, imagined and represented in its most perfected form in military academies and the battlefields of Japan's many wars.[8] The kind of men at the center of attention indicate a major shift from a primarily soldierly mode of masculinity to the masculinity of the white-collar worker, who is no longer marked by military uniforms but by business suits and whose expertise is no longer in war-making but in the pursuits of a capitalist market economy.

NEURASTHENIA: THE GLOBAL NERVOUS BREAKDOWN

The rise of neurasthenia in Meiji-era Japan became a marker for understanding how well Japan already was integrated into the modern world, as Japanese debates about its causes and consequences tied into a worldwide pattern in which the emergence of sexological discourse and the rise of empires and capitalism led to the appearance of neurasthenia in many different places. One such place was Germany. As early as 1813, a German medical doctor wrote in *Versuch über die Nervenkrankheiten (On Nervous Diseases)* that

> the most terrible consequences of this weakness and the exhaustion of nerve strength [. . .] can be observed among onanists. Most epileptics, cataleptics and morons, even the mad were onanists during their youth [. . .]. Nature penalizes masturbation even more strictly than fornication with a prostitute by syphilis.[9]

At the end of the nineteenth century, the *Handbuch der Neurasthenie* (Handbook of Neurasthenia) (1893), edited by the German physician Franz Carl Müller, contained a nearly exhaustive bibliography and thus stamped the disorder with the high-status seal of German medical science. Subsequently, the most distinguished medical men of the modern world contributed to the voluminous literature on the ailment of neurasthenia. New symptoms were added to the old ones. According to these authors' records, neurasthenics suffered from a broad spectrum of symptoms: irritability, depressive moods, abnormal fatigue, weak memory and concentration, sleep disorders, anxiety, phobias, obsessions, hallucinations, hyperaesthesia, allergies, headaches and migraine, spasms and convulsions, loss of appetite, indigestion, palpitations, nervous cardiac weakness, and sweating, as well as disorders of the sexual functions.[10]

In the United States, the neurologist George M. Beard coined the English term "neurasthenia" in 1869. Beard regarded neurasthenia as a family of disease problems long recognized by laymen and medical professionals in the

United States. Calling neurasthenia the "American disease," he believed that it was much more common in the United States, especially in the Northeast, than in Europe, as well as more common among men than among women. With the increasing interest in sexual causes of mental disorder during the late nineteenth century, Beard entertained several possible sexual sources of neurasthenia. Relating nervous exhaustion to the difficulties of modern life, he argued that neurasthenia and sexual perversion were unfortunate outcomes of progress—the dizzying growth of industry, the overcrowding of American cities, and the dissolution of moral fortitude and cultural traditions.[11]

The concept of this new disease entity developed quickly. Beard's work was accepted widely among American and European doctors, and introduced in Japan as well.[12] Considering the broad spectrum of the symptoms of neurasthenia, it is not surprising that the new concept merged and overlapped with other new disease entities, most frequently hypochondria and hysteria.[13] Jean-Martin Charcot at the Salpetrière, for example, seemed to strike a serious blow at normative masculinity in France when he extended the definition of hysteria from women to men. In 1872, he first asserted that some men subject to hysteria lacked all feminine traits. Even though they appeared to be robust men, Charcot noted, such men could become hysterical, "just like women," and this tendency was "something that [had] never entered the imagination of some people." As a rule, Charcot added, hysteria developed in men after a physical trauma, usually experienced at the workplace, while women "went hysterical" due to an overpowering emotional experience.[14] What seemed shocking for Charcot's contemporaries about his claims was that nervousness, after all, was the very opposite of the image of ideal masculinity.[15]

In Austria, the psychiatrist and sexologist Richard von Krafft-Ebing gave the diagnosis of neurasthenia a new spin. In an 1899 article on the lack of sexual feeling, he wrote that surely neurasthenia and other disturbances of the nervous system negatively affected the function of the genitals and reproductive organs.[16] He also believed that masturbation could induce neurasthenia, which in tainted individuals, could deteriorate further into homosexual perversion.[17] Leaning toward psychological explanations, his countryman Sigmund Freud attributed the cause of neurasthenia more specifically to masturbation, *coitus interruptus*, problems with the means of birth control, and the danger of venereal disease infection, all of which vitally weakened the male body's nerve force. About male hysterics, he wrote that they are "abnormally holding on to the past," and that they give in to "abnormal tendencies instead of going about their business."[18]

In 1926, Magnus Hirschfeld, one of Germany's most prominent specialists of eugenics and sexology, claimed that false ideas about the pathological character of masturbation had lost their power among medical experts. He remembered the relief he and his fellow students had felt when his teacher Wilhelm Erb announced that it was a mistake to ascribe serious illnesses such as the softening of the brain to *Ipsation* (masturbation).[19] However, in Germany and Austria, as well as in other European countries, many influential contemporaries held on to the old views. Masturbation was considered a threat to the very physical and moral fiber of the "race," and thus even supposedly progressive writers encouraged parents and teachers to do all within their power to prevent or stop the habit among their children.

Anxieties about children's sexual awakening also were at the heart of early debates about sex research and sex education in Japan, as I show below. Here I would like to emphasize that at the end of the 1920s, when Habuto's suicide was reported, neurasthenia or "a lack of nerve force" had reached significant currency in debates about the national condition in terms of physical and mental health, military strength, and Japan's potential for empire building, precisely because neurasthenia had become associated with practices of self-destruction including suicide.

Neurasthenia in the Imperial Army: Body Building and Empire Building

The modern mass military, founded in 1872, became a primary and increasingly crucial site for "body-building" efforts from the late nineteenth century onward, when the physiological male body became a central organizing principle of the nation state that was built primarily in the imperial armed forces. The imperial military was the first organization under the control of the national government to deliberately and gradually adopt several elements of a Western diet in order to improve the physique of its members.[20] The imperial armed forces were also the first organization that was drilled in the modern rules of public and personal hygiene (see also Oberländer's chapter in this book). Moreover, military-style exercise became the basis for gymnastics later introduced in schools as a fundamental tool to increase student fitness. Through these techniques, the armed forces also introduced and aggressively cultivated images of ideal masculinity and "true manhood" emerging from statistical data and averaging in the creation of modern masculinity in Japan. "Real men" were at least 1.55 m tall and somewhat heavier than the average young man. A certain lung capacity proved their fitness, and their overall health was attested to by their freedom from a number of illnesses including tuberculosis and venereal diseases, both of which were frequently associated with neurasthenia.

Being evaluated as healthy enough to serve in the imperial armed forces and particularly to be a class A recruit carried a certain prestige, not only among the ranks but back home in remote villages and towns as well.[21] Those men who did not pass the military physical or seemed unwilling to serve were in a way stripped of the prerequisites of ideal manhood. Military surgeons reported them as "simple" and "naive," documented their feeble constitution, and classified them as "lazy" and "effeminate."[22] While a positive evaluation of their bodies did not turn every young man into a soldier willing to die for the emperor and the nation, some men who were denied this seal of true manhood responded to this unfavorable classification with drastic actions including suicide.[23] Once recruited, soldiers' masculinity was constantly monitored and their health frequently checked and documented. The diagnoses of hysteria, neurasthenia, and venereal and mental diseases deemed them somewhat less than true men, as defined by the military authorities.

By the 1910s, the Japanese imperial army and navy health reports documented cases of both hysteria (*hisuterii*) and neurasthenia as separate categories among a continuously expanding list of mental diseases among soldiers. Magazines for military personnel also had begun to discuss the symptoms of and cures for neurasthenia.[24] "It is especially important," one Japanese military surgeon emphasized in 1919, "to make sure that military leaders do not suffer from neurasthenia. They have to keep their nerve until the very end of a war," he noted, "and to suffer from neurasthenia would be very dangerous in the case of military leaders who work under a lot of pressure." According to this military surgeon, "nothing new and good [could] occur without a strong body and mind."[25] The homosocial setting of the military, however, might have been a breeding ground for the affliction. Among other prominent contemporaries, Ōsugi Sakae was trained in the "ethics of warriors" at the military cadet school between 1899 and 1891, but he also engaged in what he referred to as the "vices of bushidō." A central figure in the left-wing radicalism of the early twentieth century, Ōsugi recalled in his autobiography that at the age of 13 he masturbated two or three times a day. At the military cadet school, he had been diagnosed with neurasthenia and given a two-week leave of absence at the age of 15. "I, who had been so studious," he remembered, "became a complete idler."[26] The understanding of masturbation as harmful, and of neurasthenia as an indicator of more severe ailments, also was reinforced by the fact that Ōsugi, as well as other cadets, were thrown out of the military academy for their inappropriate behavior. As protectors of national security, and representatives of a strong and healthy manhood, soldiers with diseases were particularly worrisome for the state: the military administration had to pay the cost of treatment in addition to the salaries of ill soldiers; neurasthenia was believed to lead to other, fatal illnesses such as venereal diseases and tuberculosis; and, in severe cases, patients who suffered from neurasthenia and other mental ailments were prone to suicide and seemed to offer a glimpse of the empire's fragility that had been a concern since the foundation of the modern nation state.

Throughout the modern world, military physicians especially, and public health officials more generally, noted with concern the challenges to the modern stereotype of a strong and determined masculinity. American fellow military surgeons, for example, chimed in when they found that neurasthenia was the most common malady among military men who had returned from the battlefields of World War I.[27] One medical doctor announced at the meeting of the Kansas Medical Society at Topeka in 1916 that "the rapidity with which the number of neurotics, perverts and homosexuals is increasing is appalling and that if these tendencies are not checked, their effect in the not distant future will show a decided deterioration of the race." This physician also claimed that neurasthenia occurred only in civilized peoples. In his words, the result of the malady was "a more or less perverted, weak and inefficient product [that] tends strongly to neurosis, and is unfit for marriage or parenthood [. . .]."[28] In this physician's opinion, neurasthenia occasionally affected individuals of exceptional brilliance but more often those who suffered from limited wage and uncertainty of employment, children in public schools who were crowded to the point of exhaustion, and the middle class in general.[29]

In the Japan of the 1920s, neurasthenia also had slipped decisively outside the boundaries of military medicine. Its textual representations in popular medical books and magazines were characterized by three traits. One was a distinct internationalism that had been common not only for medical theories, as Oberländer shows in this book, but for other new sciences as well. This internationalism was rooted in the quest for modernity. On the one hand, some Japanese contemporaries were critical of the unquestioned application to Japan of what they viewed as Western ideas of national health and sexual control. After all, nutrition was different, one pedagogue pointed out, as was the physical constitution of the Japanese people. In the West, another teacher suggested, people ate meat and drank beer and wine for lunch and dinner. As both meat and alcohol increased sexual desire, the threatening consequences—that is, neurasthenia—could be observed among Western youth far more often than among Japanese youth. As Morris Low points out in the Introduction, medical findings about "civilization diseases" helped confirm the modernity of Japan, but these diseases also emphasized the Japanese empire's fragility.

The first generation of Japanese psychiatrists reconceptualized *shinkei suijaku* as a particular type of nervous disease or *shinkeishitsu*. The psychiatrist Morita Shoma, for example, described the typical neurasthenic in the following words:

> [He is] a person with a particularly strong need to live a full life, perfectionist tendencies, and extreme self-consciousness. This person encounters some unpleasant event that focuses attention on a particular problem; blushing, headaches, and constipation are typical examples. He becomes quite concerned about the problem and increasingly conscious of its effects on his life. He becomes caught in a spiral of attention and sensitivity which produces a sort of obsessive self-consciousness.[30]

The causes of these weaknesses were ascribed to the rapid social transformation at the beginning of the twentieth century that seemed to challenge established notions of masculinity and manhood that were then defined largely in bodily terms. When physical differences between the Japanese and other peoples were proclaimed, for example, the "races" were hierarchically ordered according to "racial" categories that placed Japanese men below "white" men and, at least in some accounts, below other Asian men. When more and more European and American anthropologists and physicians "discovered" Japan during the late nineteenth and early twentieth centuries, they commonly described the physical condition of Japanese men in derogatory terms. European authors who were concerned about the development of the "races" were especially keen to discuss the nature and significance of the intellectual and emotive characteristics of the Japanese. Their small stature and frailty led members of the Caucasian "race" to discount them as of little or no consequence to the future achievements of mankind.[31]

Similar to the fears about military and biological decline in France, Britain, and America at the end of the nineteenth century, and Japan's eagerness to build a strong, modern army at the same time, the economic crises and the hardships of city living in Germany, China, and Japan from the late 1920s, led to a frequently voiced anxiety regarding not only physical, but also mental fitness.[32] Japanese intellectuals and, later, physicians, nutritionists, and other scientists (and charlatans), responded to this unfavorable classification in various ways, ranging from practical efforts to improve the physique of Japanese men and women using a variety of techniques informed by eugenics to far-reaching attempts at the transformation of cultural practices. Newspapers frequently reported on the physical condition of young men (and, to a lesser extent, women).[33] Hygiene and fitness programs for pupils, mothers, factory workers and white-collar employees were accompanied by an enormous amount of books and other publications on how to ensure a respectable body height and weight for growing young men.[34] An array of products and techniques that promised this very same effect was marketed in magazines and newspapers. One such advertisement promised that "Even Short Men Will Become Tall" (see figure 2.2). It could be discreetly ordered by mail order, and—according to the description in the advertisement—it lengthened not only the bone of the whole body but also the cartilage between the bones. It could be used at home in order to increase one's height and increase one's posture—thus making it a "prime invention of national interest."

NEURASTHENIA IN PEDAGOGY: MIND VERSUS BODY

Pediatrics was another field of medicine that concerned itself with neurasthenia from the time of its founding days. In May 1899, an article on the causes of "Psychological Illness among Children" that appeared in the academic journal *Jidō Kenkyū* (Pediatric Research) cited masturbation (*shuin*) as one factor leading to the occurrence of these illnesses.[35] Soon other pediatricians expressed similar views in other journals that dealt with children's development, care and health in contributions about, for example, dangers to a child's health during the growth phase, or venereal diseases in children.

In 1900, the medical doctor and distinguished historian of medicine Fujikawa Yū (1865–1940) published an article on sexual desire (*seiyoku*) in children in *Pediatric Research*, five years before Sigmund Freud caused a stir among his colleagues by publishing *Drei Abhandlungen zur Sexualtheorie* (Three Essays on the Theory of Sexuality), and before August Forel's influential *Die sexuelle Frage: Eine wissenschaftliche, psychologische, hygienische und soziologische Studie* (The Sexual Question: A Scientific, Psychological,

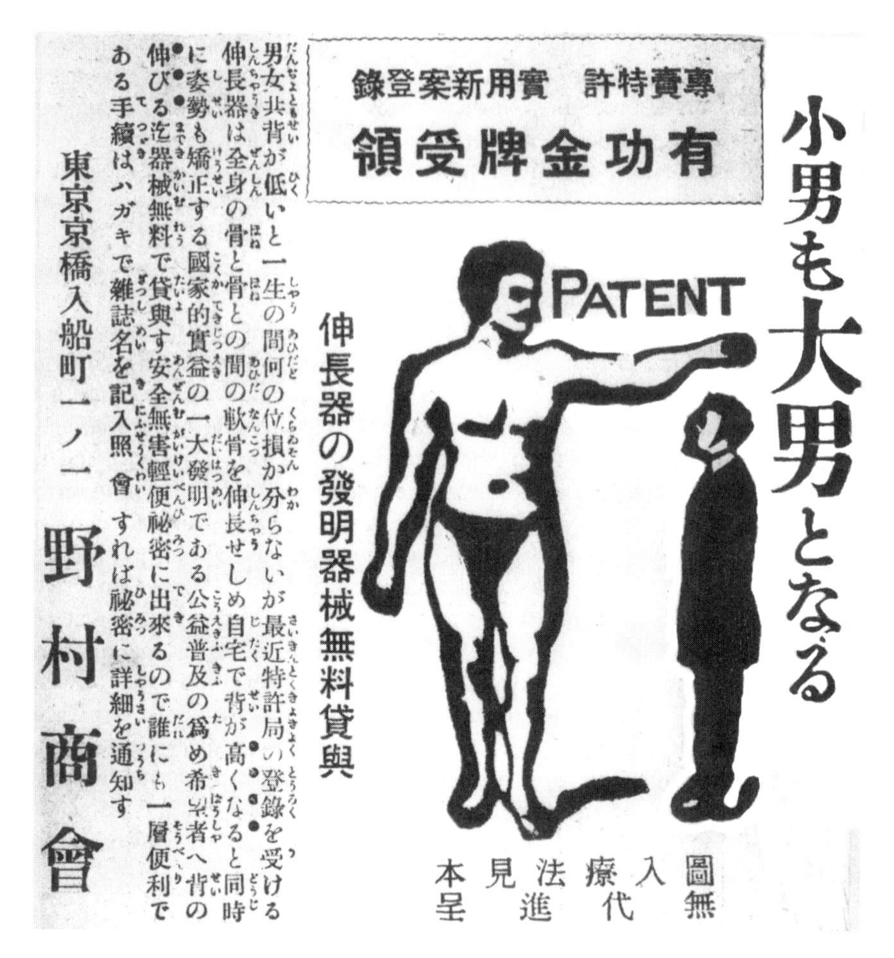

Figure 2.2 This is an advertisement for techniques to lengthen the body promising that "Even Short Men Will Become Tall."

Source: *Shinseinen* (New Youth), July 1925, p. 112.

Hygienic and Sociological Study) came out in 1905.[36] In this article and in many others to follow, Fujikawa insisted that children masturbated because they were not properly educated on sexual matters by their parents. This redefinition of neurasthenia by pediatricians marked an important new branch of inquiry that added to earlier efforts in military medicine.[37]

A few years later, the daily *Yomiuri shinbun* printed, for nearly two months, a series of articles on the necessity of sex research and sex education that were written primarily by high school and university teachers. The confusion among these writers about the age of sexual maturity was closely related to the question of masturbation, unquestionably the *béte noir* of almost every *fin de siècle* writer on human sexuality.[38] Subsequently, this debate and its spin-off in other newspapers and general women's, household, hygiene, and health magazines, all of which targeted a wide readership, triggered a nationwide discussion on the creation and dissemination of sexual knowledge. Japanese physicians, pedagogues, and bureaucrats shared their concerns with many others who worried about the future of the Japanese nation. Among their dominant convictions was the view that a lack of sexual knowledge among children and youth would not be a problem in itself if it were not for neurasthenia, which was commonly diagnosed by school physicians and generally explained as caused by "sexual immorality," that is, masturbation. School physicians rejected the claims of teachers and parents—that the weak physical constitution of many students was caused by studying too hard—and repeatedly stated that neurasthenic disturbances were caused by students' autoerotic practices. The popular consensus in Japan was that weaknesses caused by neurasthenia struck the educated upper classes in particular but potentially challenged every man's body and mind. Representations of the malady in newspapers and magazines varied. Echoing European and American ideas about the connection between the malady and intellectual capacities, neurasthenia in intellectuals was ascribed to weak nerves, psychological illnesses or simply intense intellectual work. When striking less educated men in the military, factories, and schools, however, neurasthenia was clearly, and to an increasing extent, associated with harmful sexual practices. This latter understanding of neurasthenia as an ailment resulting from certain sexual behaviors and affecting sexual functioning gradually gained ground during the early decades of the twentieth century.

Author and medical doctor Mori Ōgai embedded masturbation as a cause of neurasthenia in his autobiographic text *Wita Sekusuarisu* (Vita Sexualis). Although not a member of the naturalistic school whose representatives defended their bold thematization of sex against critics from the older generation, Ōgai was convinced of the importance of correct knowledge about sex for Japan's youth. In his novel, the narrator Kanai Shizuka describes his sexual development up to the age of nineteen, from his first childish look at erotic woodblock prints to his first visit to a courtesan. An attempt at publication in the July 1909 issue of the literary magazine *Subaru* failed. Ōgai's text was one of many to be confiscated or banned for the "stimulation of low instincts," only to be rediscovered in the 1930s as an important pedagogical document of sex education that described "sexual development in such an objective way that parents need not blush."[39]

Public health officials viewed the phenomenon in somewhat less benign terms. In 1922, for example, Ōkuma Shigenobu—a prominent politician and educator who had been appointed Minister of Domestic Affairs, Foreign Minister, and Mayor of Tokyo—presented to the participants of a conference on mental illnesses, an appeal for a law for the institutionalization of the mentally ill. His appeal clearly was dominated by the potential challenge to public health and order that the mentally ill posed. "Insanity occasionally becomes infectious," he claimed, and "this infection can be terrible, spreading ceaselessly among the people." Ōkuma also suggested that a society, or even a state, can eventually become morbid, and he named Russia as an example of a nation

Yasuda, described their findings countless times in articles and speeches in an explicit attempt to liberate the masses from incorrect sexual beliefs. Similar to the Austrian psychiatrist Richard von Krafft-Ebing, the Englishman Havelock Ellis, and his Japanese colleague Habuto Eiji, Yamamoto received many letters from the readers of his articles on sexual behavior and the audience of his public speeches on sex education. In his replies, he repeatedly asserted that masturbation was "normal" and not to be worried about too much.[50] Yamamoto claimed that based on his surveys, there was very little correlation between class and sexual behavior in a broader sense. More than half, or 631, of the men he had surveyed about their sexual experiences had had their first sexual intercourse before the age of 18. About half of those who had graduated from a middle school had chosen a prostitute for their first partner. This trend was also the case for a third of those who had graduated from a high school or a university.[51] Other surveys carried out in Japan as well as in European countries reflected similar results.[52]

Neurasthenia and the Power of Capitalism

Ironically, just as Yamamoto and other doctors, biologists, and sexologists began to publicly dismiss theories about the supposedly harmful consequences of masturbation as nonsense, neurasthenia was becoming recognized in the popular media as one of the most common ailments, and the pharmaceutical industry jumped at the opportunity to put new products for treatment on the market. Even more ironically, neurasthenia was perceived by many as a side effect of an urban, capitalist, consumerist society with its different patterns of work and production; yet the cure was more consumption—of new medicines. During an era of intense militarization, empire building, warmaking, and aggressive pronatal policies, the symptomology of neurasthenia became more and more reduced to sexual dysfunction, and the earlier, more diverse interpretations gave way to anxieties about the decline of Japanese men's virility, and perhaps about the lurking fall of the Japanese empire. The pharmaceutical industry responded to the disease that gained stature, hand in hand, with the emergence of a consumer society, by recommending greater consumption—perhaps also because whether they perceived neurasthenia as the result of a misled sexual desire, the lack of knowledge on correct sexual behavior or the underdevelopment of the genitalia, many medical doctors, government officials, and pharmaceutical experts increasingly linked individual masculinity with male sexual potency and imperialist efforts.

Popular texts and the explosion of potency-enhancing pharmaceutical products responded to long-harbored anxieties that Japanese men's bodies were not as well developed as those of other peoples and that their sexual potency was being challenged by non-procreational practices. Family and household magazines, women's magazines, and general magazines, as well as newspapers, began to discuss symptoms and cures. Advertisements for clinics explicitly announced that examinations and treatments were on offer for the following: neurasthenia, problems concerning the sexual organs, and chronic gonorrhea, thus suggesting that someone who suffered from neurasthenia could be infected by more serious venereal diseases such as gonorrhea as well.[53] In 1930, Abe Isoo, for example, the founder of the first Japanese socialist party and a fervent activist in the Purity Society (Kakuseikai), which fought for the abolition of prostitution, submitted a report to the journal *Tsūzoku igaku* (Popular Medicine) about his recovery from neurasthenia in order to encourage other sufferers to consult a particular hospital in Tokyo. In the article, Abe reported that he had suffered from acute neurasthenia after a serious illness. He could not sleep and became forgetful. Just when his pride in having been healthy for 60 years began to crumble, Abe wrote, he decided to consult the Yamashita Kōryō Clinic in Tokyo and was healed by a 50-time treatment there.[54]

Advertisements that were printed in various media promoted a variety of hormone products mostly for men.[55] In 1927, *Popular Medicine*, for example, ran an advertisement for the hormonal product Tokkapin.[56] As pictured in the advertisement, a young man assuming a Superman-like pose, with overly long legs, holds up a packet of the product. He wears a suit and a bowtie and stands on a whole pile of Tokkapin packets. Through his legs and behind him, the reader sees the smoking chimney of a factory in front of which a few *rikisha*-men are on their way to serve customers. The advertisement features the young, successful, middle-class man who can—in addition to all of his other achievements—strengthen the functions of the genitalia and increase his energy in general, simply by taking Tokkapin. According to the text, Tokkapin also could cure a number of other ailments including impotence, premature ejaculation, nocturnal pollution, decline of sexual strength, senility, loss of staying power, hysteria, insomnia, amnesia, anemia, wrong nutrition and, of course, neurasthenia.

In 1933, the same magazine featured a hormone tablet for the treatment of "sexual defects" and "incomplete development of the genitalia." The text of the advertisement claimed that injections and other complicated methods of treatment had become unnecessary. Instead, it announced that the medical world now welcomed and highly praised this new method of treatment for "sexual neurasthenia," incomplete development of the sexual organs, atrichia, frigidity or apathy, and other disorders.[57] A test sample could be ordered by sending a meager two sen in postage stamps to the Japan Society for Popular Medicine in Osaka.

Another advertisement for hormone tablets sold at the Shisandō Pharmacy in Tokyo[58] promised to heal a decline in sexual desire, premature ejaculation, nocturnal pollution, and a number of other disorders. According to the explanation in the advertisement, the substance for the tablets was scientifically extracted from the genital glands of healthy bulls. The advertisement claimed that, according to results from recent research in internal medicine, the hormone extracted from the sex glands had proved to be highly effective in the treatment of "sexual neurasthenia." Advertised as a "hormone product," the price of 2.5 yen for 50 tablets and 7 yen for 150 made the product a luxury item, at a time when one could buy a book for 1 yen and the average urban middle-class household monthly income was about 60 yen.

In yet another advertisement, hormone tablets were described as especially successful for the treatment of premature ejaculation.[59] In the advertisement, physicians described their successful treatment, typically addressing married men who were unable to enjoy a satisfactory sex life. According to the text of the advertisement, the tablets did not cause dependency, but one can imagine that those who could afford it might have taken more than recommended and thus spent even more money on them. Hormone treatment was, in any case, a luxury for most of them, as the quantity of hormone tablets to be used for about 22 days cost 2.5 yen, for one month it was 4.5 yen, and for two months it was 7.80 yen. Hormone products for the treatment of neurasthenia and the improvement of sexual potency, as well as general physical strength, clearly were intended to appeal to men, and they did so in a very peculiar fashion by inciting sexual desire and simultaneously restricting its scope to heterosexual practice. Experts, identified as medical doctors, explained the "scientific methods" used in the production of the medication. An increase in sexual desire and a cure for sexual malfunctioning were promised.

While sexual intercourse was not mentioned explicitly, many of these advertisements featured a woman's face or other parts of a woman's body. These illustrations of female body parts seem to suggest why nocturnal pollution, premature ejaculation, neurasthenic ailments, or a lack or decrease of sexual desire should be cured. It is the photographs of women's faces that represent a sexual counterpart and reinforce what is implicitly suggested in the text—that nonreproductive sex is a waste of energy. A lifted skirt, a pair of bare legs or a smiling female face seem to be intended to stimulate sexual desire in the male reader, on the one hand, and at the same time emphasize that sexual desire (which these images might provoke) and pleasure had to be shared with a woman. It is significant to note that regardless of the prevalence of the presence of parts of a naked female body, her sexuality is absent as she only exists as the potential object of male desire. Although some readers suggest in their statements that they tried the products because they were concerned about not being able to satisfy their wives, female sexuality was merely utilized for the stimulation of male readers' fantasies and desires.

The post–World War II era provided a new framework for a significantly transformed manhood, as well as a reconfigured relationship between the individual male body and the nation at peace.[60] Neurasthenia survived the war almost unharmed, but it was stripped of its sexual and imperialist connotations in the process. It reappeared in various popular medical books for home use, typically in the category "neurological diseases" either under its old name "*shinkei suijaku*" or renamed "*shinkeishō*" (nervous disorder or neurosis) (see figure 2.3). According to the self-help book *Manseibyō no*

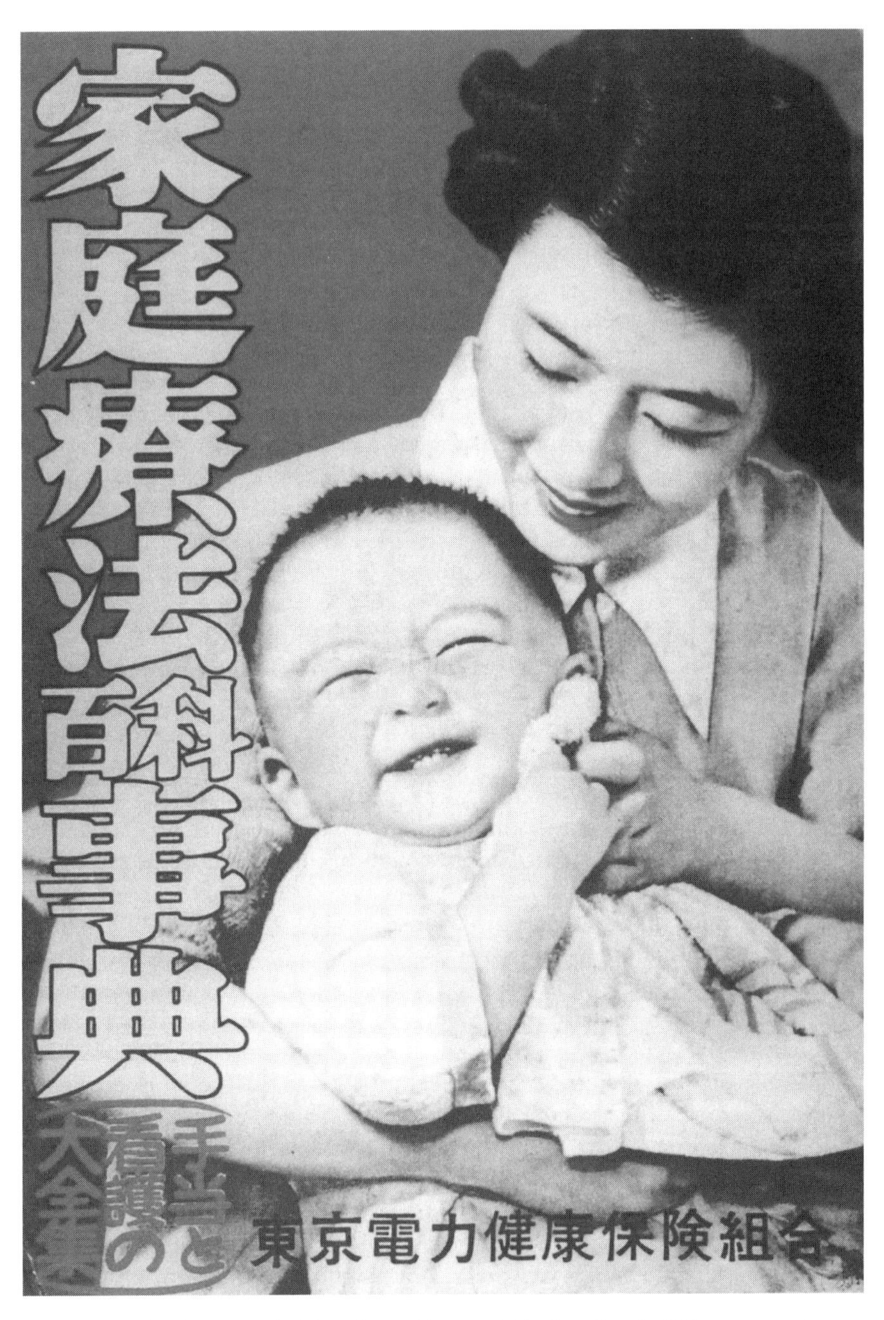

Figure 2.3 The home medical handbook *Katei ryōhō hyakka jiten* (1952) was one of many medical handbooks that were available to the readers of primarily women's magazines and other members of the middle and upper classes. It described neurasthenia as a (male) phenomenon primarily due to overwork, and differentiated it from (female) hysteria and other ailments of the nervous system.

katei ryōhō (House Remedies for Chronic Diseases), a supplement to the November 1949 issue of the women's magazine *Shufu no tomo* (Women's Companion), for example, neurasthenia was an ailment that could easily occur in people who were overworked, had too many concerns, or suffered from a lack of sleep.[61] No potency-enhancing product would appear as prominently in popular media again until Viagra was legalized and hit the Japanese market in 1999—just in time to reassure Japanese men (and other men all over the world) who were in the midst of a severe economic crisis, of their sexual potency despite the sharp decline of their economic and social power.[62]

CONCLUSION

From the late nineteenth to the mid-twentieth centuries, Japanese scholars and practitioners in medicine, pedagogues, psychologists, hygienists, as well as philosophers and bureaucrats, developed a new understanding of the programming of the Japanese body and, through it, of the entire population. They increasingly attempted to make use of scientific knowledge in order to form well-functioning and well-regulated bodies that would constitute a better and more modern nation. "Neurasthenia," among other illnesses, seemed to put these reform attempts at risk. Stirring anxieties about both individual manhood and the power and stability of an empire in the making, its symptomology tied together the ability to overcome the challenges of modern life, the necessity of scientifically informed sexual behavior, and the capacity to launch imperialist actions.

During the early 1920s, sex reformers attempted to educate the masses on correct sexual behavior, and—among other new ideas—began to vehemently oppose the view that masturbation could result in neurasthenia, social turmoil, or even a revolution. Instead, they urged the public that masturbation was harmless and even part of a normal sex life. By the early 1930s, however, popular print media, pharmacies, and medical doctors joined in an otherwise unlikely alliance in their search for new ways of increasing profits. They recognized the commercial value of advertising sexual potency and the possibilities of increasing it. Advertising sexual health and strength was widely tolerated by the censors during the 1930s, most probably because it fed into the increasing attempts by authorities, leading intellectuals and social movements to build a physically and mentally strong population, enhance its procreative potential, and thus program a nation fit for war. Japanese imperialism needed not truthful knowledge but sexually and militarily potent men. Hence, in modern Japan, the creation of sexual knowledge and the stigmatization of certain kinds of sexual behavior, as well as the promotion of others, were intrinsically intertwined with politics. By the mid-twentieth century, this configuration had culminated in a rhetoric relating to the militarization of sexuality and the sexualization of the military, ultimately followed by a sense of failure of modern (militarist) manhood.

Notes

* I am grateful to the MedHeads at the University of California at Berkeley and Warwick Anderson in particular for their comments on an earlier version of this chapter. Michael Bourdaghs' critique has been invaluable for broadening my perspective. I also would like to thank Hiromi Mizuno and the participants in the workshop on "Sex and the Politics of Desire: Japan" at the University of Minnesota in April 2002. Research and writing were facilitated greatly by the University of California President's Fellowship in the Humanities and a Committee of Research Grant from the University of California at Santa Barbara.

1. Yokoyama Tetsuo, "Seigaku no taika Habuto hakushi shinkei suijaku ni taoru," *Tsūzoku igaku* 1929, *7* (10): 1–4.
2. See, e.g., *Enjū satsuyō* (1631) and *Yōjōkun* (1714); both cited in Shimizu Masaru, *Nihon no seigaku jishi* (Tokyo: Kawade Shobō, 1989), pp. 199–206 and 246.
3. As Thomas Laqueur has shown in *Solitary Sex: A Cultural History of Masturbation* (Cambridge: Zone Books, 2004), concerns about masturbation had been a global phenomenon.
4. The term "disease entity" was coined in 1935 by the microbiologist and sociologist of science Ludwik Fleck in order to differentiate between the different stages of a disease and to highlight the various transformations in the understanding of a disease depending on the availability of medical and other scientific knowledge. Similar to Fleck's study of the various causes, symptoms and healing techniques of syphilis in Europe between the fifteenth and the twentieth centuries, the causes of neurasthenia in *fin de siècle* Japan were described in diverse and complex ways.
5. This view was expressed in the daily newspaper *Yomiuri shinbun* in articles published throughout September and October 1908.
6. Matsumoto Shizuo, "Seishin suijaku to shuin (onanii)," *Tsūzoku igaku* 1937, *15* (3): 98–100; Mukō Gunji, "Seiyoku mondai o shitei ni oshifuru no rigai 1," *Yomiuri shinbun*, September 1, 1908: 5.
7. Scholars of Japan have begun to write a history of masculinity only recently; most notable among them are Hikosaka Tai, *Dansei shinwa* (Komichi Shobō, 1991); Itō Kimio, *"Otokorashisa" no yukue: Dansei bunka no bunka shakaigaku* (Shinyōsha, 1993); Inoue Teruko, Ueno Chizuko, and Ehara Yumiko, eds., *Danseigaku* (Iwanami Shoten, 1995); and Taga Futoshi, *Dansei to jendā keisei* (Tōyōkan Shuppansha, 2001). Recent approaches to the topic in the United States include George L. Mosse's examination of *The Image of Man: The Creation of Modern Masculinity* (Oxford University Press, 1996); David D. Gilmore's *Manhood in the Making: Cultural Concepts of Masculinity* (Yale University Press, 1990); and R.W. Connell's *Masculinities* (University of California Press, 1995).
8. The main military conflicts in which Japan was involved include the Sino-Japanese War of 1894–1895, the Russo-Japanese War of 1904–1905, and the conflict provoked by the Kwantung Army in 1931, which resulted in the establishment of Manchukuo, and the beginning of a full-blown war with China in 1937 that culminated in 1941 in the attack on Pearl Harbor and the Pacific War.
9. Cited in Magnus Hirschfeld, *Geschlechtskunde auf Grund dreißigjähriger Forschung und Erfahrung 1* (Stuttgart: Julius Püttmann Verlagsbuchhandlung, 1926), p. 287.
10. Andreas Hill, " 'May the doctor advise extramarital intercourse?': Medical Debates on Sexual Abstinence in Germany, c. 1900," in *Sexual Knowledge,*

Matsumura Noriaki, Hirono Yoshiyuki, and Matsubara Yōko, "Fujikawa Yū: Pioneer of the History of Medicine in Japan," *Historia Scientiarum* 1998, *8* (2): 157–171; Fujikawa Hideo, ed., *Fujikawa Yū chosakushū* (Kyōto: Shibunkyaku shuppan, 1982); and Fujikawa Yū sensei kankōkai, ed., *Fujikawa Yū sensei* (Tokyo: Daikūsha, 1988).

37. For literary authors around 1900, neurasthenia was a largely personal affair. In their writings, neurasthenia primarily appeared as a foil for contemplations on an intellectually and emotionally demanding modern lifestyle. In Natsume Sōseki's novels, neurasthenia is commonly used to describe and explain the flaws of the leading characters. Neighbors ascribe neurasthenia to the leading character in *Wagahai wa neko de aru (I Am a Cat)* as well as in *Kusamakura* (The Three-Cornered World). Natsume Sōseki suffered from the disease himself and obviously attempted to increase sympathy and understanding for other victims of the ailment through his sympathetic description of "neurasthenics" (*shinkei suijaku-sha*) in the novels he wrote in Tokyo after he had returned from England in 1903. He also might have been influenced by the aforementioned renowned psychiatrist Morita Shoma, with whom he was acquainted. See Takahashi Masao, "Sōseki bungaku ni okeru naoshi. 'Shinkei suijaku'-sha no rikai to kyūsai," *Nihon byōsekigaku zasshi* 1996, *52*: 30–36.
38. R.P. Neuman, "The Sexual Question and Social Democracy in Imperial Germany," *Journal of Social History* 1974, *7* (3): 271–286, on pp. 272–273.
39. The censorship of *Wita sekusuarisu* was reported in *Taiyō* January 1, 1908, cited in Jay Rubin, *Injurious to Public Morals: Writers and the Meiji State* (Seattle and London: University of Washington Press, 1984), pp. 21–22. For the rehabilitation of the text by sexologists, see Takasugi Saburō, "Ryōsho no suisen," *Seikagaku kenkyū* 1936, *1* (5): 78.
40. Nakatani Yoji, "Relationship of Mental Health Legislation to the Perception of Insanity at the Turn of the 20th Century in Japan," unpublished manuscript, 1995, p. 15.
41. There is a huge volume of literature on neurasthenia and its place in the history of sexual knowledge. For the European history see the following authors: Karl Braun, *Die Krankheit Onania. Körperangst und die Anfänge moderner Sexualität im 18. Jahrhundert* (Frankfurt am Main: Campus Verlag, 1995); Arnold Davidson, "Sex and the Emergence of Sexuality," *Critical Inquiry* 1987, *14* (1): 16–48; and Neuman, "The Sexual Question," For Europe, see Roy Porter and Leslie Hall, *The Facts of Life: The Creation of Sexual Knowledge in Britain, 1650–1950* (New Haven and London: Yale University Press, 1995); Robert A. Nye, "The History of Sexuality in Context: National Sexological Traditions," *Science in Context* 1991, *4* (2): 387–406; and Peter Weingart, Jürgen Kroll and Kurt Bayertz, *Rasse, Blut und Gene: Geschichte der Eugenik und Rassenhygiene in Deutschland* (Frankfurt am Main: Suhrkamp Verlag, 1992), pp. 108–113. For the discussion of sexological ideas in Russia see Laura Engelstein, *The Keys to Happiness: Sex and the Search for Modernity in Fin-de-siècle Russia* (Ithaca and London: Cornell University Press, 1992); and for China, see Kleinman, *Social Origin*; and Frank Dikötter, *Sex, Culture and Modernity in China: Medical Science and the Construction of Sexual Identities in the Early Republican Period* (London: C. Hurst & Co., 1995).
42. *Fujo shinbun*, June 19, 1921, cited in Furukawa Makoto, "Ren'ai to seiyoku no daisan teikoku," *Gendai shisō* 1993, *21* (7): 114.
43. *Taiyō*, January 1, 1908; cited in Rubin, *Injurious to Public Morals*, pp. 121–122.

44. Ibid., October 1, 1908, p. 122.
45. Mukō Gunji, "Seiyoku mondai to kongō kyōiku," *Yomiuri shinbun*, September 10, 1908, p. 5.
46. Yoshida Kumaji, "Seiyoku mondai o shitei ni oshifuru no rigai 3," *Yomiuri shinbun,* October 11, 1908, p. 5.
47. Inagaki Suematsu, "Seiyoku mondai to kongō kyōiku," *Yomiuri shinbun* September 10, 1908, p. 5.
48. Mukō Gunji, "Seiyoku mondai," p. 5.
49. The first survey of sexual behavior in boys and girls was published in 1949; Asayama Shin'ichi, *Sei no kiroku. Sengo Nihonjin no seikōdō o kagakuteki ni chōsa shita shiryō ni motozuku* (Ōsaka: Rokugatsusha, 1949).
50. Yamamoto Senji, *Yamamoto Senji zenshū: Daiikkan jinsei seibutsugaku*, ed., Sasaki Toshiji (Tōkyō: Sekibunsha, 1979), pp. 104–105.
51. Okamoto Kazuhiko, "Taishū no gaku toshite no seikagaku no tenkai," *Gendai Seikyōiku Kenkyū* 1983, *14*: 108–118.
52. See, e.g., Sugita Naogeki, "Seibyō to seishinbyō no kankei," *Kakusei* 1924, *14* (8): 17–19; Yoshii Kaneoka, "Eiseijō kara mitaru kōshō mondai," *Kakusei* 1940, *30* (1): 38–39; and Alfred Blaschko, "Eiseijō yori kōshō seido o ronzu," *Kakusei* 1914, *4* (3): 5–11.
53. *Tsūzoku igaku*, October 1930, *10* (10): 175.
54. Abe Isoo, "Shinkei suijaku nanshōsha ni," *Tsūzoku igaku* 1930, *10* (8): 152.
55. Matsumoto Shizuo, "Seishin suijaku to shuin (onanii)," *Tsūzoku igaku* 1937, *15* (3): 98.
56. *Tsūzoku igaku* 1927, *5* (1): advertisement section.
57. *Tsūzoku igaku* 1933, *11* (10): 156.
58. *Tsūzoku igaku* 1933, 11 (8): 124.
59. *Tsūzoku igaku* 1938, *16* (1): 114.
60. I discuss different aspects of the complex transformation of military masculinity in the post–World War II era in the following articles: " 'Nur nicht kampflos aufgeben!' Die Geschlechter der japanischen Armee" (" 'Don't Give up without a Fight!' The Genders of the Japanese Military"), in *Gender und Militär: Internationale Erfahrungen und Perspektiven*, ed. Christine Eifler and Ruth Seifert (Berlin: Ulrike Helmer Verlag/Heinrich Böll Stiftung, 2004), pp. 155–187; "Männer, Tauben und Kirschblüten: Zur kollektiven Gedächtnisproduktion in Militärmuseen" ("Men, Doves and Cherry Blossoms: On the Production of a Collective Memory in Military Museums"), in *Innovationen in der Japanforschung* (Münster: LIT Verlag, 2004), ed. Roland Domenig, Susanne Formanek, and Wolfram Manzenreiter, pp. 1–27.
61. The women's magazine *Shufu no tomo* had been particularly active in providing medical handbooks to its readers; in these handbooks reputable medical doctors explained the causes, nature, and healing techniques for a great variety of ailments. A wartime example of such a handbook is the *Musume to tsuma to haha no eisei dokuhon* (Hygiene reader for daughters, wives, and mothers), published in 1937. Other organizations produced similar self-help books, such as the *Katei ryōhō hyakka jiten* (Encyclopedia of House Medicine) that was published by the Tokyo branch of the Union for Electricity and Medical Insurance in 1952.
62. The significance of Viagra as a marker of the transformation of the social, economic, and gender order since the 1990s is discussed in the epilogue of Sabine Frühstück, *Colonizing Sex: Sexology and Social Control in Modern Japan* (Berkeley: University of California Press, 2004).

3

The Female Body and Eugenic Thought in Meiji Japan

Sumiko Otsubo

Introduction

Japan is renowned for its "selective adaptation of ideas and institutions."[1] This chapter deals with one example, the transplantation and domestication of "eugenics."[2] "Eugenics" is a term coined in 1883 by British scientist Francis Galton to describe the notion that human genetic stock could be improved by controlling heredity. The boundary between the "fit" who were encouraged to reproduce, and the "unfit" often coincided with boundaries of "race," gender, and class. It is thus intriguing to ask why some Japanese chose to adopt and adhere to the Western science of eugenics, even though it seemed to prescribe inferior status to the Japanese in a white-dominated international "racial" hierarchy. In the late nineteenth and early twentieth centuries, Japanese leaders, aspiring to make Japan capable of competing with industrial and "civilized" Western nations, launched comprehensive modernization programs. Scientists were among those who eagerly participated in this process of "building a new era."[3] In this context, eugenics can be seen as a "biological" approach to this far-reaching modernization plan.

In this chapter, I explore the eugenic thought of physiologist Ōsawa Kenji (1852–1927), and the ways in which scientific authorities were employed in efforts to apply eugenic policies to society. Ōsawa was one of the first scientists to systematically "medicalize" race improvement discourse,[4] which had been dominated by nonmedical professionals, including Fukuzawa Yukichi. Ōsawa's ideas were pivotal in the history of eugenics in Japan because he emphasized the female body as a strategic site in which constitutional improvement of the Japanese "race" could be made. He saw that women's bodies could be eugenically controlled by marriage, and advocated the exchange of prenuptial health certificates, prepared by qualified physicians. In other words, he medicalized one of life's most important events, marriage.[5] Moreover, he allowed feminists, educators, and social reformers, particularly temperance activists, to appropriate his scientific authority, hoping that they, in return, would help him put his eugenics proposals into practice.

As a professor of medicine at Tokyo University,[6] Ōsawa, of course, was a state employee. By examining these two-way interactions, I demonstrate the complex relationships between agents of the state and private citizens involved in eugenic policy formulation. We can also observe a pattern of transplantation of a foreign idea. Ōsawa emphasized indigenous customs, including arranged marriage, as conducive to the Japanese adoption of eugenics. This conscious mobilization of local practices as "traditions" is a fairly typical response to Western-inspired modernity in Meiji Japan.

EARLY LIFE OF ŌSAWA KENJI

Ōsawa Kenji was born to the family of a Shinto priest in 1852. His name by birth was Ōbayashi Ukonji. As a child, Ukonji was adopted by Ōsawa Genryū, a medical doctor who had been trained in European medicine in Nagasaki and was serving the local Mikawa domain lord in modern day Aichi prefecture. Before leaving for Edo (present day Tokyo) in 1866, Kenji received a samurai education—he studied Confucian classics at a domain school—since the Ōsawa belonged to the warrior class. Political disturbances, which led to the breakdown of the traditional Tokugawa order and the establishment of the modern Meiji government (the 1868 Meiji Restoration), interrupted Ōsawa's study at the Shogunal Institute of European Medicine, in the capital city of Edo. Yet, he managed to resume training at the same school after it was taken over by the new Meiji government. In 1870, Meiji leaders sent Ōsawa, along with 13 other students, to Europe. He pursued his study in medicine at Berlin University. There he took Hermann von Helmholtz's physics and Emil du Bois-Reymond's physiology classes. Although the government wanted him to study pharmacology, du Bois-Reymond's class further stimulated Ōsawa's interest in physiology, which he had developed while reading imported textbooks in Tokyo. Because of a government policy change, he was called home in 1874 before completing his doctoral study. After a few years of teaching physics and physiology as an instructor at Tokyo University, he resigned the post to finish his postgraduate study in Europe, 1878–1882. This time he specialized in physiology at Strassburg University.[7] He chose Strassburg because the hired foreign (*oyatoi gaikokujin*) physiologist at Tokyo University, Ernst Tiegel, recommended his former teachers. At Strassburg, he studied closely with the medical chemist Felix Hoppe-Seyler and physiologist Friedlich Leopold Geoltz. Ōsawa's dissertation research was a neurophysiological study concerned with transmission in dogs' spinal cords.

Upon returning home in 1882, 30-year-old Ōsawa was immediately appointed Professor of Physiology of the Faculty of Medicine at Tokyo University. He replaced Tiegel as holder of the chair of physiology. As his interest in hygiene grew, Ōsawa also taught that subject during this time. He also organized an interdisciplinary medical study group whose members, mostly professors, exchanged new knowledge acquired by reading the most recent Western journals in their respective fields. Even after his retirement from Tokyo in 1915, he continued to publish many articles and books concerning a wide array of topics such as diet, digestion, excretion, hunger, development of various senses, reproduction, heredity, anesthesia, drinking, and sexology before his death in 1927. During his tenure, Ōsawa assumed numerous important positions both within and outside the university, including deanship of the Faculty of Medicine, and membership in the House of Peers by Imperial decree (*chokusen kizokuin giin*).[8] Some of his articles advocated meat eating (formerly proscribed by Buddhist teachings) and improvement of the human bodily constitution. These articles reflected the optimistic belief that conscious effort could ameliorate the Japanese body's appearance (i.e., size and shape) as well as capability (i.e., speed, power, and endurance). As such, they embodied the Meiji reform spirit applied to customs and morals.

TAKAHASHI YOSHIO'S RACE IMPROVEMENT THEORY

Beginning in the 1880s, the theory of evolution captivated the thinking of Meiji intellectuals.[9] The theory served as a scientific endorsement for the notion that the body was open to biological reconstruction; and in this context, the message of eugenics was attractive. Indeed, as early as 1881, two years before Francis Galton coined the term "eugenics," the prominent promoter of Western ideas, educator, and journalist Fukuzawa Yukichi, commented on Francis Galton's study concerning inheritance of talents.[10] In 1884, Fukuzawa's protégé, Takahashi Yoshio, published Japan's first book on race betterment, *Nihon jinshu kairyōron* [On the improvement of the Japanese race]. Here Takahashi discussed how to improve the Japanese race and proposed different approaches to achieve this goal. He supported his arguments with the theories of many Western scholars, including those of Francis Galton. But he did so without referring to the newly invented term, "eugenics," which means "well-birth science" in Greek. In addition to emphasizing the reform of physical education, clothing, diet, and housing, Takahashi suggested intermarriage between the Japanese and "whites." After presenting the statistical data of physical size among different nationals, he showed that an average Japanese was shorter and lighter than an average (white) Westerner; and the cranial size of the Mongoloid was smaller than that of the Caucasoid, implying that the former's mental capacity might be inferior to the latter's. As a quick remedy to the "undesirable" Japanese body, Takahashi suggested the "crossbreeding" of the two "races."

Takahashi's proposal provoked nationalistic reactions from some of Japan's leading men of learning, including a professor of philosophy at Tokyo University, Inoue Tetsujirō, and the president of the same institution, Katō Hiroyuki. The debate over the Japanese version of "whitening" took place in the context of Japan's aspiration for equality with the West.[11] Along with the establishment of tariff autonomy and elimination of extraterritorial jurisdiction, the Japanese were discussing whether or not they should allow Westerners to live with the Japanese outside of the designated treaty ports where Westerners had been confined. While pro-Westernization advocates, including Takahashi, supported mixed residence, which would result in the increase of "crossbreeding," others, such as Katō and Inoue, were more cautious. Thus the latter group argued that, from a social Darwinist perspective, the Japanese, as the less "civilized" people at least for the moment, were likely to lose out to more advanced Westerners both commercially as well as biologically. Katō was particularly alarmed by the possible disappearance of the "pure" ("*junsui naru*") Japanese race. The popularity of the pro-mixed residence arguments peaked around the mid-1880s but declined and came under severe criticism during the reactionary intellectual climate of the late 1880s.[12]

This mixed marriage/residence debate clearly showed that some Japanese felt they were "racially" inferior to Westerners. Though many were anxious to "improve" the Japanese body, Takahashi's approach, which denied the preservation of the existing Japanese identity, was adamantly rejected. But Takahashi's radical proposal spurred more serious discussion about race improvement based on the idea of controlling heredity through marriage. While environmental approaches such as better nutrition, clothing, and living conditions coincided with (middle class) women's expanding sphere of influence at home, the notion of reproductive race betterment truly brought to the fore the role of women in this important reform movement.[13]

Except for Erwin von Baelz, the Tokyo University professor who taught internal medicine and pathology between 1873 and 1902, few medical or biological experts were actively involved in the debate. However, it is unlikely that the controversy started by Takahashi Yoshio went unnoticed by Ōsawa, who had just returned from Germany. First, the president of the university for which Ōsawa worked was a major participant in the debate. President Katō expressed his view on a high profile occasion, a speech at the Tokyo Academy, and his response to Takahashi's view of the Katō speech was printed in *Tōyō gakugei zasshi*, a respected journal modeled after Britain's *Nature*.[14] Second, Takahashi's race improvement approach touched on Ōsawa's own research subjects of diet and reproduction.

Bodily Improvement and the Female Body

Ōsawa Kenji began writing about marriage in 1890. However, it was only in 1904 that Ōsawa began to advocate bodily improvement (*taishitsu kairyō*) through selective breeding, and stress the significance of the female body in this process. The years between the mid-1880s and 1904 saw several changes worthy of attention. First, the German cytologist August Weismann (University of Freiburg) in 1883 provided evidence antagonistic to the notion that acquired characteristics are inheritable. Weismann argued that germ plasm (sperm and egg cell nuclei) could not be affected by the environment, and was completely isolated from somatic (or body) cells, which could be. Weismann's doctrine of the "continuity of the germ plasm" and the Mendelian laws, which were rediscovered in 1900, reinforced each other in explaining the phenomenon of heredity. Heredity now became the subject of intensive scrutiny by biologists.

Second, the Japanese government began incorporating the official gender ideology of "good wife, wise mother" into the curriculum of secondary schools for girls after Japan's military victory over China in 1895. Third, educator Naruse Jinzō, dedicated to promoting higher education for women, elaborated on the official ideal of womanhood, and argued that women's physical and mental quality would have a direct impact on future generations of Japanese. Thus, in his 1895 book, *Joshi kyōiku* [Women's education], Naruse explained that a scientific approach to producing mentally, physically, and morally "fit" women would be crucial for Japan's nation-building. Naruse's far-reaching fund-raising campaigns, which drew support from prominent politicians of the day, finally paid off when his brainchild, Japan Women's College (*Nihon Joshi Daigakkō*), was established in 1901.

Meanwhile the government had founded a teacher training college for women in 1875, and a few other private "women's colleges" opened in 1900. They specialized in English, medicine, or art education. These colleges started with only a few faculty members, including the founders, and several students. The Japan Women's College, also a private institution, greatly differed from other colleges for women in that, from the beginning, the College was able to provide a well-diversified and well-balanced liberal arts education. It was made possible because Naruse enthusiastically recruited about 50 qualified teachers including Ōsawa Kenji, who taught physiology at the College between 1901 and 1921.[15]

Fourth, after five years of study in animal and human anatomy at the University of Freiburg, which was a stronghold of the scientification of eugenic theories,[16] Ōsawa's adopted son, Gakutarō (1863–1920), returned home with his German wife, Julia Meyer, to assume a professorship at Tokyo University in 1898. In addition to his scholarly works, Gakutarō also wrote essays on Japanese women.[17] The earlier debate over mixed marriage began to bear personal implications for Ōsawa Kenji. Considering that Kenji began teaching at the Japan Women's College, it was likely that women's issues became a frequent topic of discussion between father and son. Furthermore, while it was certainly likely that the father received up-to-date biological and medical theories from Germany through the son who took August Weismann's course, among others, the older Ōsawa had a first-hand opportunity to get reacquainted with European biomedical communities in 1901. He presented a paper at the International Congress of Physiology in Turin, Italy, and another at the International Congress of Zoology at Berlin.[18] Professionalization of eugenics in Europe did not immediately follow Galton's invention of the term "eugenics" in the 1880s. Rather, the professionalization, which replaced the preceding "liberal and secular cultural movement," began taking shape about the time that Ōsawa revisited Berlin.

This led to the institutionalization of eugenics as evidenced by the founding of the Racial Hygiene Society in Berlin in 1905 and the Eugenics Education Society in London in 1907.[19]

Ōsawa Kenji's 1904 work, *Shakaiteki eisei taishitsu kairyōron* [On the improvement of human bodily constitution from a social hygienic perspective], reflected the scientific and intellectual developments of the preceding two decades. He utilized newly available statistics in Europe and Japan, and identified what kind of diseases and problems would be harmful. His use of the term *shakai eisei*, its focus on the degeneration of "civilized" people in a domestic context, as well as many of his statistics and examples from works by German theorists[20] seem to indicate that the book drew inspiration from the contemporary German notion of *Sozialhygiene*.[21] Indeed, despite its title containing the word "*kairyō*" (improvement), the book was more concerned with the prevention of degeneration than betterment per se. He was convinced that a civilization, after reaching maturity, tends to decline because of racial degeneration.[22] In the second half of the nineteenth century, many European specialists, especially in criminal anthropology and psychiatry, noted the paradoxical nature of civilization—"science and economic progress might be the catalyst of, as much as the defense against, physical and social pathology."[23] Like them, Ōsawa believed the mechanism of natural selection (the survival of the fittest) no longer worked in a modern society because modern medical care artificially extended the lives of the weak, who were naturally "unfit" for survival, and helped them produce offspring with "unfit" genes. Moreover, it was believed that "[m]oral decadence, chronic diseases like tuberculosis, venereal diseases and alcoholism, crime and deviant social behaviour—which included merely having two children or less," frequently observed in "civilized" societies, were considered factors contributing to racial degeneration.[24]

Overtly concerned with the possible decline of the human race by the breakdown of natural selection, Ōsawa classified people into four general categories: those who were fit to have intercourse (*kōsetsu tekisha*); those who were not (*kōsetsu futekisha*); those who were fit to reproduce (*seishoku tekisha*); and those who were not (*seishoku futekisha*). Certain diseases such as tuberculosis, leprosy, syphilis, and gonorrhea would spread through intercourse. While the carriers of diseases would ruin the health of their sexual partners, people who were too young or too old would harm their own bodies. Intercourse would cause pain for those with sexual organs that were underdeveloped or had stopped functioning properly, or became deformed after menopause.[25] Although this applied to both men and women, Ōsawa noted that women were subject to more restrictions. Women should refrain from having sex when in periods of menstruation, puerperium, or lactation, because, for example, if a nursing woman had intercourse, her body might stop lactating. During pregnancy, women should not have sex, or at least reduce the frequency, Ōsawa explained, because copulation might induce miscarriage or inflict other types of damage on the fetus.[26]

In his discussion on reproductive fitness, Ōsawa further differentiated the female body from the male body. The birth of a healthy child requires three components: a perfect sperm cell, a flawless egg cell, and a mother's robust body.[27] Women, associated with two of the three, obviously had a greater role in reproduction. Concerns with the overall quality of children, which had serious implications for the future of the nation, brought medical attention to the female body. Ōsawa concluded that undesirable intercourse and reproduction could be controlled by education, laws, and contraceptive methods such as condoms and spermicide.[28] To avoid the spousal and trans-generational spread of diseases, Ōsawa proposed the prenuptial exchange of health certificates.[29] This proposal attests to the remarkably current nature of Ōsawa's knowledge. In the same year (1904) when Ōsawa wrote this, it was recorded that Francis Galton's paper provoked discussion on the desirability of prospective bridegrooms to obtain medically certified documents in England. In Germany, the League for the Protection of Mothers and Sexual Reform and the Monist League, both founded in 1905, advocated marriage health certificates, which led to a legislative effort during World War I.[30]

Ōsawa's arguments regarding intercourse and reproductive fitness could be applied to any other "civilized" society. They were more universal than nationalistic. Yet, he did touch on a few local conditions, peculiar to Japan. He observed that many middle- to upper-class women in Europe regrettably avoided breast-feeding for aesthetic reasons and instead used alternative artificial (meaning nonhuman) milk, which made their children's constitution more likely to be inferior. The use of artificial milk was not yet widespread in Japan.[31] Another favorable custom (among upper-class Japanese) was premarital detective investigation to check for the presence of diseases such as tuberculosis and leprosy in the prospective spouse's family. These diseases were considered contaminants to the family blood and lineage. As the Tokugawa orders restricting the change of hereditary professions and residences were lifted, conducting this kind of detective work became steadily more difficult.[32]

MEDICALIZATION OF MARRIAGE AND THE FEMALE BODY

Three years after he published the *Taishitsu kairyōron*, Ōsawa wrote *Seirigakujō yori mitaru fujin no honbun* [The duty of women from the physiological point of view] in 1908. As the title indicates, in this work he intensified his attention to women and their bodies as a crucial object for successful implementation of the bodily improvement theory. This rather brief book[33] was soon expanded into a 584 page volume, *Tsūzoku kekkon shinsetsu* [Popular new theory on marriage]. It was originally printed as a series entitled "*Kōfuku naru kekkonhō*" ["Ways to ensure a happy marriage"] in the *Hōchi* newspaper and was soon revised and published in book form in 1909. In her article "Marriage, the Newspaper Business, and the Nation-State," Kathryn Ragsdale examines popular romance fiction featuring married female protagonists, a new genre known as the domestic novel, which became common between the Sino- and Russo-Japanese wars (1895–1904). She points out that, in the late Meiji period, newspaper editors discovered the importance of female readership for expanding their sales, and began printing the serialized domestic novel. Ōsawa's "Kōfuku naru kekkonhō" appeared in the *Hōchi* in this context. The *Hōchi* and its detective agency that would investigate the prospective husband's and wife's individual and family background, promoted Ōsawa's scientific gospel concerning marriage. The agency sold Ōsawa's marriage guidebook in its office and stressed its own capacity to ferret out health information. Thus, Ōsawa's books on bodily improvement, mark not only the scientification of eugenic theories and medicalization of marriage, but also the commercialization and popularization of race improvement through matrimony.[34]

In the introduction of the *Tsūzoku kekkon shinsetsu*, Ōsawa stated that the primary objective of living organisms was to perpetuate their species through reproduction. Humans were no exception. They would achieve this goal by marriages that would determine the fortunes and misfortunes of individuals and families as well as the rise and decline of Japan. Many young men and women had no idea of what was at stake in choosing mates. Especially when they made hasty decisions driven by the temporary passions of love, their marriages tended to result in various health problems among family members, including children. Ōsawa wrote this book as a scientific guide for marriage in the modern era.[35]

Elsewhere he also told readers that marriage could harm the body in different ways. For instance, diseases such as venereal diseases and tuberculosis would be transmitted between husband and wife. The serious nature of health problems was obvious because the mental and physical conditions of both parents would influence the quality of children.[36] Although the title and the introduction were not particularly gender-specific, the main text of the book clearly intended to offer insights mainly for young single educated women who were preparing to marry. Objectification of the female body was justified because women were more likely to become victims of ailments caused by bad marriages. In a patriarchal society like Japan, a (middle to upper class) woman upon marriage was generally expected to leave her own home and move into the husband's. The new wife had to deal with unfamiliar customs, including having sex, in the new environment surrounded by unsympathetic strangers. Moreover, she had to submit to the authority of the husband as well as to those of the father- and mother-in-law. Together with the belief that the female nervous system was more sensitive, the greater stress level on women tended to have a negative impact on women's health. Women were more susceptible to health problems when they went through pregnancy and birth.[37] Compared with nineteenth-century French medical doctor Gustave Le Bon, who had maintained that women were both psychologically and physiologically sensitive to civil strife in general,[38] Ōsawa's emphasis on the connection between the Japanese family system and the mental and physical stress on women is notable. Like European theorists, Ōsawa considered women "a crucial agent of degeneration either . . . by bringing new pathological cases into the world or . . . by failing to reproduce in sufficient quantity healthy children for the nation."[39] However, because the readers of his book were expected to be middle to upper class women, Ōsawa focused more on women's ability to control the quality of offspring than the danger presented by "unfit" mothers. In the first section of the *Tsūzoku kekkon shinsetsu*, Ōsawa stated:

> [E]ducation for women should aim at producing morally, mentally, and physically "fit" women. These . . . fit women serve as the most important mechanism to improve our racial stock . . . or the Yamato *minzoku* [Japanese race]. . . . [T]he improvement of humans requires men and women of superior quality, which I will explain in the following chapters. . . . To be frank from the standpoint of racial stock improvement, it is actually desirable that those women without much education and cultivation—thus they were close to animals—give birth to few children. However, we would like women of superior quality with education and cultivation to have as many children as possible.[40]

The boundary between "fit" and "unfit" women as described above is defined by whether or not they had received higher education, which was deemed to guarantee morally, mentally, and physically improved women.[41] The improved qualities acquired by a mother's education would then be transmitted to their children genetically. This optimistic two-step approach closely resonated with educator Naruse Jinzō's race improvement view presented in the 1890s.

To support his view, Ōsawa discussed various genetic theories, which made their appearance after the rediscovery of Mendel's Laws. Unlike August Weismann, who thought only egg and sperm cell nuclei could transmit parents' characteristics to their children, Ōsawa believed that egg and sperm cells as a whole (each cell was made up of a cell nucleus [*kaku*] and protoplasm [*genkeishitsu*]) served to transmit inheritable qualities. Ōsawa accepted Weismann's theory of the continuity of germ plasm. The Japanese physiologist, however, disagreed with the German biologist in that the former believed protoplasm *could* be affected by environment. Thus, Ōsawa postulated that environment (e.g., nutrition or substances) would affect protoplasm, which would then influence cell nuclei. In short, his conviction that one could improve the female body through physical exercise and better nutrition and hygiene was based on this genetic view.[42] His distinction between congenitalness (*senten*) and posteriority (*kōten*) reflected his theoretical understanding of heredity. He specified that while congenitalness meant transmission of parents' characteristics to their children before fertilization, posteriority meant the transmission of characteristics after fertilization.

Ōsawa believed that congenitalness depended on two basic patterns. First, when human reproductive organs were developing, egg and sperm cells (both cell nuclei *and* protoplasm in Ōsawa's understanding) were susceptible to changes in the nervous system, which controlled reproductive functions. Thus, malnutrition, immaturity, senility, or excessive drinking would affect egg and sperm cells. Second, after they completed their development but before fertilization, their egg and sperm cells could be influenced by the parent's consumption of various substances, including alcohol.[43] Ōsawa's interest in physiology was concerned with nutrition, growth, reproduction, motion, senses, and mental activities, and thus made him attentive to constitution or race improvement theory.[44] His theoretical understanding catered to the assumption that physiology could improve an inferior physique and serve to better the human species. Ōsawa, a trained physiologist, was thus responsible for medicalizing race or constitution improvement theory. This is extremely important considering the fact that another physiologist Nagai Hisomu (1876–1957), Ōsawa's successor at Tokyo, later emerged as Japan's most prominent eugenicist. He popularized eugenic theories, promoted eugenic research and policies, organized scholarly and popular eugenic associations, and lobbied for the enactment of the 1940 National Eugenics Law.[45] An equally significant point is that Ōsawa's optimistic view of heredity allowed the Japanese to believe in efforts to improve their bodies within the framework of science. Healthy lifestyles led by young adults, whose egg or sperm cells were already mature and waiting for fertilization, amounted to quality control over the nation's population. This rejection of outright biological determinism, albeit not uniquely Japanese, explains why some Japanese embraced eugenics. Their sensitivity about their apparent physical inferiority, evidenced in the earlier mixed marriage debate, fueled interest in eugenics, rather than rejection of it.

Another characteristic of this book was its emphasis on the Yamato *minzoku*, as distinct from Westerners. This element hardly existed in the 1904 *Taishitsu kairyōron*, which discussed bodily improvement in a more universal, and biological, but less nationalistic, and cultural sense. In his *Tsūzoku kekkon shinsetsu*, Ōsawa called Japan the country of gods (*Shinkoku*) and argued that the family system and the samurai spirit (*bushidō*) were essential in shaping the Japanese.[46] Mental activity was part of physiology, as Ōsawa understood it. Thus, the samurai tradition of *bushidō*, based on self-sacrificing loyalty to the master, became a physiological subject. In addition, the purity (*junketsu*) of the *minzoku*, ancestor worship, Confucianism, and Buddhism formed the uniquely Japanese altruistic spirit. He also stressed that the Japanese, as descendants of an unbroken line of rulers, were united through the worship of their common ancestors, the imperial clan.

Women's concept of *bushidō*, spiritual conscience or *reinōteki ryōshin*, was seen by Ōsawa as suppressing their self-serving, animal-like sexual desire (*dōbutsuteki seiyoku*) for the good of others.[47] As such, it needed to be preserved. He discouraged "fit" women from pursuing Western-inspired free love (*jiyū ren'ai*).[48] He associated free love with primal sexual instinct and condemned it as a force destructive to the family state; he implicitly defended the framework of the traditional upper class custom of arranged marriage. Its mechanism of choosing the most suitable spouse for one's daughter or son was definitely compatible with the notion of race improvement through controlling heredity. Modern changes in the traditional institution of marriage created a new standard of spousal selection. Now "biological" fitness was added as the most important consideration to ensure happy arranged marriages and the continuation of a family line. Healthy couples without hereditary or infectious diseases were likely to produce physically "fit" children. Educated parents were expected to stay away from possible health hazards. Ōsawa did not embrace everything Western, nor did he dismiss everything traditionally Japanese. He found certain indigenous practices useful as foundations for "transplanting" eugenics in Japan. His eugenics represented a "hybrid" reinterpretation of Western and Japanese cultures, not a mere transfer of original eugenics to a new environment.

One sees a striking difference in tone between the *Taishitsu kairyōron* (1904) and the *Kekkon shinsetsu* (1909). While the former discussed bodily improvement in general, the latter presented a much more "racialized" view in the context of social Darwinist, imperialist competition. When "yellow" Japan defeated "white" Russia in 1905, many Asian and African peoples of color colonized by the "white" Europeans and Americans were inspired by this victory. At the same time, the erosion of their "racial" supremacy alarmed the "white" imperialists (the "yellow scare").[49] The heightened interest in "racial" competition and the military spirit that existed immediately after the Russo-Japanese war were evident in the *Kekkon shinsetsu*. Ōsawa's constitution/race improvement writings thus represent a dramatic shift from the early Meiji "catch-up" spirit of "the reform of customs and morals" to the late Meiji mentality of nationalism, which resulted from Japan's emergence as a colonial power competing against Western rivals.[50]

Interactions between the Medical Authorities and Social Reformers

Ōsawa developed his "scientific" race improvement ideas and promoted them in his interactions with individuals outside the medical profession in the Meiji (1868–1912) and Taishō (1912–1926) periods. First, employment at Naruse Jinzō's Japan Women's College gave him direct and regular opportunities to speak to female college students about the transgenerational implications of their bodies. Ōsawa's awareness of women's potentially instrumental role in diffusing "modern" hygienic concepts led him to give occasional lectures at meetings of the Greater Japan Private Women's Hygienic Association even before the turn of the century. Yet, as noted, his involvement in the discussions about bodily improvement through marriage became much more active after he started work at the College in 1901. After its founding, Naruse focused on expanding the curriculum and operations. When a new educational law opened the way for some qualified colleges (*senmon gakkō*) to become universities, Naruse aspired to elevate his college. To convince the public that his school deserved to become a university, he announced a school expansion plan in 1917. He proposed to add a faculty of medicine, including a department of race improvement (*jinshu kairyō gakka*), to the Japan Women's College. Both in school and public lectures, as well as publications of articles and books by Ōsawa greatly contributed to legitimizing the college's claim that it was already committed to improving the nation's genetic quality.

One of the students who responded to the calls of Ōsawa and Naruse was Hiratsuka Raichō, who would become the most prominent Japanese feminist in the twentieth century. She majored in home economics at the Japan Women's College between 1903 and 1906. This home economics program was likely the best science education available for women in the country at that time.[51] When women were still denied even participation in political meetings, Hiratsuka attempted to establish Japan's first eugenic law in 1919. Her daring proposal sought to prevent men (but not women) infected with venereal diseases from getting married. Existing scholarly writing often emphasizes that Hiratsuka was inspired by Swedish feminist Ellen Key's motherhood ideology, to which she was exposed in the 1910s.[52] Yet, one should take into account the fact that Hiratsuka was a student of Ōsawa's when he first suggested marriage restriction against the venereally diseased in his 1904 *Taishitsu kairyōron*. Furthermore, like Ōsawa, she also promoted the use of prenuptial health certificates. Criticized for her gender-specific approach to male marriageability as outlined in the petition draft, Hiratsuka, for political reasons, needed to revise the draft and looked to Ōsawa for advice. This is wholly appropriate, for, in his 1909 *Kekkon shinsetsu*, he had delineated three approaches to marriage: individualistic (*kojinteki*), racial (*jinshuteki*), and social (*shakaiteki*).[53]

In the first type, individuals were responsible for choosing their marriage partners (love marriage). Ōsawa was opposed to this because young people tended to make the most important decision of their lives driven by sexual desire, disregarding crucial conditions such as physical fitness and education levels. The opposite extreme was racial marriage. Its primary objective was to improve race (*jinshu kairyō*). It could be achieved when the state intervened in the lives of individuals: the "fit" were allowed to marry, but the "unfit" were not. Ōsawa admitted that he had supported this approach in *Taishitsu kairyōron* (1904) and classified the "insane," "imbecilic," those with syphilis or gonorrhea, alcoholism, epilepsy, or genetic diseases, as "unfit." Although he used the term "racial (*jinshuteki*)," his writing did not explicitly imply competition between the Japanese "race (*minzoku*)" and other "races (*minzoku*)." By "*jinshu*," it seems that he meant state supervision and intervention.[54]

By 1909, however, he had abandoned the racial approach and had begun advocating the social approach, which was a combination of the individualistic and the racial. The racial approach was ineffective when people reproduced outside the marriage institution. In the social strategy, individual freedom was to be respected as much as possible and individuals were responsible for whatever decision they made. However, at times, the state needed to interfere with individual freedom in order to improve national health (*kokumin no kenkō*). What Ōsawa had in mind was for the state to order physicians to prepare health certificates for those about to marry. If any mental or physical problem were detected, the state would prohibit their marriage and sterilize them so that they would not do harm to the state by producing undesirable offspring. Ōsawa compared the state with a human body. If one were starving, he or she would lose fat, muscles, glands, and bones while maintaining the basic weight of the brain, spinal cord, and heart, which were essential for survival. Like this "natural" mechanism of the body, the welfare of the state (i.e., brain and heart) had to be prioritized over the freedom of individuals (i.e., fat, muscles, etc.).[55] Ōsawa thus identified himself as a "statist" (*kokka shugi o hōzuru mono*).[56]

Hiratsuka's original proposal called for men to present a document guaranteeing their venereal-disease-free status to their prospective brides-to-be before marriage. Sanctions would be imposed on men who got married while concealing a venereal disease. After incorporating Ōsawa's advice, which was obviously based on the social approach explained in *Kekkon shinsetsu*, Hiratsuka expanded her proposed law to cover not only regular marriage, but also *de facto* marriage (*jijitsukon*), to minimize the births of children to parents carrying venereal diseases.[57]

The collaboration between Ōsawa and Hiratsuka, who was known as a frivolous, radical woman tainted by scandals, including a suicide attempt, alcohol consumption at bars, and public debate on such taboo topics as abortion and sexuality, was a strange one. After all, Hiratsuka was a feminist who questioned patriarchal societal norms and such authorities as the state and the father. She was against any arranged marriage that subordinated women's interests to family welfare. Moreover, she chose to have children without marrying her partner. Ōsawa, on the other hand, advocated modified arranged marriage for eugenic purposes and viewed illegitimate children negatively.[58] As long as Hiratsuka used eugenic reasoning, however, Ōsawa found her legislative initiative useful for implementing his race improvement policy. And he was willing to endorse her project in the public media.[59] At the same time, Ōsawa's support was a valuable asset for Hiratsuka. His prestige as a Tokyo University professor, and his medical expertise, legitimized her effort to protect middle-class women from diseases carried by their potential mates.

In a society where the notion of "men's predominance over women" (*danson johi*) dominated, and the belief that only men's characters would affect children's because women merely served as "borrowed wombs" was generally accepted, Ōsawa's view was quite revolutionary and attractive to Hiratsuka. Furthermore, Ōsawa urged that single women be informed so that they could choose their future husbands and produce biologically desirable children, an important task for nation-building. In other words, he encouraged "fit" women to pursue postsecondary education and assume an active and assertive part in marriage decisionmaking. Considering that opportunities for higher education then were virtually monopolized by men, and middle class women were seen as virtuous if they displayed signs of submissiveness, obedience, and docility, his eugenic ideas were potentially instrumental in redefining women's narrowly prescribed role.

One of the members of the House of Representatives who introduced Hiratsuka's petition in 1920 was Nemoto Shō. He was another of the social reformers who collaborated with Ōsawa. Nemoto was an American trained temperance activist in Japan. When Ōsawa's lecture on the degenerative harm of alcohol and preventive measures was printed in the *Hōchi* newspaper in 1907, Nemoto, editor of the temperance magazine *Kuni no hikari* [Light of our land], was quick to reprint the article to justify the temperance claim of using "scientific authority."[60] Beginning in 1898, Nemoto was elected to parliament ten times. Between 1901 and 1922, he submitted a bill to restrict minors from drinking alcohol 19 times. Although the House of Representatives had begun approving the temperance bill in 1908, the Peers kept rejecting it.[61] During this time of frustration, Ōsawa, himself a member of the House of Peers, gave a speech in support of the temperance bill; the physiologist cautioned the reluctant members of the House of Peers in 1910 that unrestricted drinking would have a negative impact on the Japanese state (*kokka*), people (*jinmin*), and race (*jinshu*).[62]

Ōsawa Kenji's relationship with Naruse Jinzō, Hiratsuka Raichō, and Nemoto Shō can be seen as part of an ongoing pattern. Like Hiratsuka, who wanted to improve the well-being of women, Christian social reformers hoped to reduce the misery caused by addiction to alcohol. Likewise, Naruse wished to promote higher education for women. They advocated eugenics as a strategy for legitimizing their causes and saw alliances with Ōsawa, who had scientific authority, as beneficial. Ōsawa, too, found the collaboration with social activists helpful in converting his eugenic plan into reality. Although he was exposed to Western values through his study abroad and scientific inquiries, Ōsawa was a politically conservative statist who upheld distinctly Japanese "traditions" such as the imperial institution, family system, arranged marriage, and the way of warriors, especially after the Russo-Japanese war. He believed in state intervention into people's everyday lives and people's cooperation with the state, which would result in "racial" well-being. He was neither a liberal Christian nor an enthusiastic feminist. He never attempted to open prestigious Tokyo University to female students nor did he support the upgrading of women's colleges to universities. In spite of rather opposing ideologies, the man of medicine and the social reformers decided to work together for practical reasons.

In addition to illuminating the eugenic appeal to people with a broad range of social, political, and religious views at the time, this essay challenges the common perception of the relationship between the state and the people. Many historians contend that, because Japan, compared with some Western nation-states, started its modernization relatively late, the government took charge of active industry-building by training experts and allocating financial resources instead of waiting for private businesses to evolve. This strong government leadership is said to have been accepted because of the traditional prestige and authority associated with the official sector (*kan*). Many observe that the same top-down policymaking structure has defined modern Japanese society.[63] Many Marxist scholars have drawn attention to how people have been marginalized by eugenic laws. Some examine men and women judged as mentally or physically unfit who were considered for such negative eugenic measures as sterilization and quarantine.[64] And others find that women's bodies became the targets of state control.[65] These studies tend to portray the government as the agent objectifying and victimizing ordinary people's bodies.

As far as eugenic legislative efforts were concerned, however, prior to the enactment of the 1940 National Eugenics Law, private individuals such as feminists and Christian social reformers started many movements. They wanted the government to restrict people's bodies. Naruse Jinzō's case was not a legislative effort; but he wanted the government, more precisely Ministry of Education officials, to see that his college was qualified to attain university status and used the potential utility of women in conjunction with the new science of eugenics to try to achieve his goal. Ōsawa can be seen as part of the Government since he was employed by a state university and served as a member of the House of Peers. Yet, even he was unable to legalize his eugenic policies and sought private activists' organized support. Well into the 1930s, the government was attracted by the general eugenic message of improving the quality of the Japanese population, but many officials remained decidedly unenthusiastic about actually establishing eugenic policies. This was so because they were difficult to implement, their effects were uncertain, and above all, they were long-term investments (taking generations to get results). Japan could not afford spending its limited resources in the face of other problems that required immediate actions. Christian temperance activist Nemoto Shō too was a member of the House of Representatives.[66] Even though the lower house was a part of the state legislature, there were many instances in which the upper house and the lower house, as well as the Cabinet and the lower house, had conflicting views that were difficult to reconcile. In fact, Representatives promoting eugenic bills in the 1930s once lamented that few bills proposed by the lower house were ever enacted. Bureaucrats formulated the majority of laws.[67] Contrary to the popular image that the state monolithically and eagerly imposed the eugenic laws on the people, various eugenic legislative movements can also be seen as private citizens' efforts to convince a reluctant "state" to take control of Japanese bodies. To negotiate with the unwilling "state" more effectively, social reformers with varied agendas strategically and pragmatically mobilized the medical and scientific authority, which Ōsawa represented. Ōsawa's willingness to work with the social reformers came from his understanding that a combination of state intervention and voluntarism (or what he called the "social" approach) would be the best way to counteract degeneration in the modern era.

CONCLUSION

Examination of Ōsawa Kenji's prescriptive eugenic writings and involvement in social causes illuminates why a physiologist became involved in the medicalization of constitution/race improvement theory in Meiji Japan. His physiological interest in diet and reproduction, and a national obsession for reforming customs and morals led him to look into the science of race improvement. His interpretation of heredity allowed a greater role for physical exercise and learning in controlling population quality. He rejected strict biological determinism that disregarded the impact of physical training and education. Especially because the Japanese were anxious to correct their self-defined physical inferiority, eugenics attracted the attention of some Japanese. Ōsawa's training in physiology had much to do with his early leadership in bodily improvement movements. Ōsawa's basic medical (*kiso igaku*) research interest bordered biology (zoology) and medicine, as we have seen in his dissertation on dogs' spinal cords and his participation in the zoological conference in Berlin. Historian of science, Suzuki Zenji, saw that there was a relative lack of interest in eugenics among Japanese biologists. While the first generation of Japanese geneticists, mostly working on rice and silkworms, operated in the framework of the faculty of agriculture and were not funded to extend their research to humans, no other medical specialists were able to conduct sophisticated experimental research using human subjects either.[68] Only in the 1910s did biologists and medical experts begin to actively participate in discussions on eugenics.

This chapter also shows Ōsawa's crucial role in diffusing race improvement theory by medicalizing an important life event, namely marriage. His understanding of emerging genetic and evolution theories led him to notice the validity of the female body in race improvement. His "hybrid" eugenics, which included such elements as the promotion of the Japanese family system and arranged marriage, and emphasis on the home as women's sphere of influence, were compatible with the patriarchal values, invented as authoritative "traditions" by the Meiji officials and intellectuals.[69] At the same time, his ideas attracted pragmatic feminists such as Hiratsuka Raichō because he explained scientifically that women were at least equally responsible for determining the characteristics of offspring.[70]

Despite the fact that Ōsawa's eugenic thought contained oppressive patriarchal values, from which feminists were struggling to emancipate themselves, Hiratsuka Raichō sought advice from her former professor. And, even though Hiratsuka questioned values he believed in, Ōsawa endorsed Hiratsuka's eugenic marriage legislative movement. He saw the feminist proposal as scientifically sound and good for the nation. The collaboration between Ōsawa and Hiratsuka was similar to that between Ōsawa and others such as Naruse Jinzō and Nemoto Shō. First, individuals with different political and social visions came together to advance toward their immediate goals. Second, such an alliance represented the private citizens' active agency in influencing state policymaking. What was distinct about the Ōsawa–Hiratsuka coalition, however, was that it revealed two remarkably different interpretations regarding the significance of the female body. For Ōsawa the female body was an object to be controlled by the (male) authorities. For Hiratsuka and other women, the female body served as a bargaining chip for negotiation and a source of empowerment.

NOTES

1. I would like to thank James Bartholomew, Kevin Doak, Margaret Lock, Morris Low, Matsubara Yōko, Lawrence Sitcawich, Sharon Traweek, Yuki Terazawa, and

Rumi Yasutake for generously assisting me during the course of this research. The quote is from Mark B. Adams, "Toward a Comparative History of Eugenics," in *The Wellborn Science: Eugenics in Germany, France, Brazil, and Russia*, ed. Mark B. Adams (New York: Oxford University Press, 1990), pp. 225–226.

2. For the hierarchy of center and periphery of scientific knowledge production, see Nancy Stepan, *"The Hour of Eugenics": Race, Gender, and Nation in Latin America* (Ithaca: Cornell University Press, 1991), p. 3; Hiroshige Tetsu, *Kagaku to rekishi* (Tokyo: Misuzu Shobō, 1965), pp. 103–105; Sharon Traweek, *Beamtimes and Lifetimes: The World of High Energy Physicists* (Cambridge, Mass.: Harvard University Press, 1988); and Morris Fraser Low, "The Butterfly and the Frigate: Social Studies of Science in Japan," *Social Studies of Science* 1989, *19*: 313–342. For the concepts of transplantation, domestication, and translation, see Joseph J. Tobin, "Introduction: Domesticating the West," in *Re-Made in Japan: Everyday Life and Consumer Taste in a Changing Society*, ed. Joseph J. Tobin (New Haven, Conn.: Yale University Press, 1992), 1–41, on p. 4; and Tessa Morris-Suzuki, "The Great Translation: Traditional and Modern Science in Japan's Industrialization," *Historia Scientiarum* 1995, *5* (2): 103–116.
3. This was the assessment of the Japanese historian of science Yoshida Mitsukuni, quoted in James R. Bartholomew, *The Formation of Science in Japan: Building a Research Tradition* (New Haven, Conn.: Yale University Press, 1989), p. 4.
4. For existing work that touches on Ōsawa's eugenic ideas, see Suzuki Zenji, *Nihon no yūseigaku: Sono shisō to undō no kiseki* (Tokyo: Sankyō Shuppan, 1983), p. 92; and Saitoh Hikaru, "*Chiiku taiiku iden kyōikuron* o kangaeru: Nihon yūseigakushi no hitokoma," *Kyōto Seika Daigaku kiyō* 1993, *5*: 168–178, on pp. 171–173. Neither Saitoh nor Suzuki suggest that Ōsawa paid a special attention to the female body. Though she analyzes him more as a sexologist than eugenicist, Sabine Frühstück, however, notes that Ōzawa (Ōsawa) Kenji identified "chastity, women's participation in the workforce outside the home, and birth control" as the most important "sexual problems" in 1920. See her *Colonizing Sex: Sexology and Social Control in Modern Japan* (Berkeley: University of California Press, 2003), p. 104.
5. "Medicalization of life" refers to medical professionals' attempt to bring various events, behaviors, and problems into their sphere by diagnosing them as pathological. This means the creation of a new market because previously nonmedical matters are transformed into something that required health scientific treatment and care. See Margaret Lock, "Ambiguities of Aging: Japanese Experience and Perceptions of Menopause," *Culture, Medicine and Psychiatry* 1986, *10*: 23–46.
6. As Japan improved and expanded its higher education system, modern day Tokyo University went through numerous organizational changes. Consequently, it was renamed several times. To avoid unnecessary confusion, I use "Tokyo University" throughout this chapter.
7. At the time, Strassburg (Fr. Strassbourg), a city in modern-day French province Alsace-Lorraine (G. Elsass-Lothringen), belonged to the German Empire.
8. About Ōsawa's life, see his memoir, Ōsawa Kenji, *Tōei chūgo*, ed. Nagai Hisomu (Tokyo: Kyōrinsha, 1928). See also K.R. Iseki, ed., *Who's Who Hakushi in Great Japan 1888–1922* (alternative title: Iseki Kurō, ed., *Dai Nihon hakushiroku*), Vol. 2 (Tokyo: Hattensha, 1925), pp. 4–5, 27–28 (in English) and 4–5, 26–27 (in Japanese); Koichi Uchiyama and Chandler McC. Brooks, "Kenji Osawa, a Pioneer Physiologist of Japan," *Journal of the History of Medicine and Allied Sciences* 1965, *20*: 277–279; Sakagami Katsuya, ed., *Gekidō no Nihon seijishi*, Vol. 1, *Meiji Taishō*

Shōwa rekidai kokkai giin shiroku (Tokyo: Asaka Shobō, 1979), p. 958; and Nihon Seirigaku Kyōshitsushi Henshū Iinkai, ed., *Nihon seirigaku kyōshitsushi*, Vol. 1 (Tokyo: Nihon Seirigakkai, 1983), pp. 272–277.

9. Masao Watanabe, *The Japanese and Western Science*, trans. Otto Theodor Benfey (Philadelphia: University of Pennsylvania Press, 1990), p. 79.
10. Fukuzawa Yukichi, "Jiji shōgen," in *Fukuzawa Yukichi zenshū*, ed. Keiō Gijuku, Vol. 5 (1881; reprint, Tokyo: Iwanami Shoten, 1959), pp. 225–231. Galton published *Hereditary Genius* in 1869.
11. For "whitening" and "whiteness" discussion in China, Brazil, and Japan, see Sakamoto Hiroko, "Ren'ai shinsei to minzoku kairyō no 'kagaku': Goshi shinbunka disukōsu to shite no yūsei shisō," *Shisō* 1998, *894*:4–34, on p. 7; Stepan, "*The Hour of Eugenics*," pp. 154–156; and Morris Low, "The Japanese Nation in Evolution: W.E. Griffis, Hybridity and Whiteness of the Japanese Race," *History and Anthropology* 1999, *11* (2–3): 203–234.
12. See Takahashi Yoshio, "Nihon jinshu kairyō ron," in *Meiji bunka shiryō sōsho*, Vol. 6, *Shakai mondai hen*, ed. Kaji Ryūichi (1884; reprint, Tokyo: Kazama Shobō, 1961), pp. 15–55. For studies on this subject, see Suzuki, *Nihon no yūseigaku*, pp. 32–44; and Fujino Yutaka, "Kindai Nihon to yūsei shisō no juyō," in *Nihon fashizumu to yūsei shisō* (Kyoto: Kamogawa Shoten, 1998), pp. 371–394. In English, see Hiroshi Unoura, "Samurai Darwinism: Hiroyuki Katō and the Reception of Darwin's Theory in Modern Japan from the 1880s to the 1900s," *History and Anthropology* 1999, *11* (2–3): 235–255.
13. See Fukuzawa Yukichi, "Nihon fujinron," in *Fukuzawa Yukichi zenshū*, Vol. 5, ed. Keiō Gijuku (1886; reprint, Tokyo: Iwanami Shoten, 1959), pp. 447–474; Sugihara Naoko, "Fukuzawa Yukichi no joseiron ni okeru paradaimu tenkan," *Ningen bunka kenkyū nenpō* 1991, *15*: 219–229; and Fujino, "Kindai Nihon to yūsei shisō no juyō," pp. 386–392.
14. Later Ōsawa wrote an article analyzing the lefthandedness of Katō Hiroyuki in 1899. Ōsawa also noted that Katō was involved in the mixed marriage debate a decade and some years earlier, see Ōsawa Kenji, *Tsūzoku kekkon shinsetsu*, (Tokyo: Ōkura Shoten, 1909), p. 245.
15. Sumiko Otsubo and James R. Bartholomew, "Eugenics in Japan: Some Ironies of Modernity, 1883–1945," *Science in Context* 1998, *11* (3–4): 545–565, on pp. 549–552.
16. Paul Weindling, *Health, Race, and German Politics between National Unification and Nazism 1870–1945* (Cambridge: Cambridge University Press, 1989), pp. 96–101.
17. Iseki, *Who's Who Hakushi in Great Japan 1888–1922*, Vol. 2, pp. 27–28 (in English) and 26–27 (in Japanese).
18. Iseki, *Who's Who Hakushi in Great Japan 1888–1922*, Vol. 2, pp. 4–5 (in English) and 4–5 (in Japanese) and Nihon Seirigaku Kyōshitsushi Henshū Iinkai, *Nihon seirigaku kyōshitsushi*, Vol. 1, pp. 272–277. While his physiological paper delivered in Turin was concerned with lefthandedness, his zoological study presented in Berlin was about the collective move of a kind of fish, *Itome*. Upon returning from Europe, he gave a talk on his observation during the trip at a meeting of the Greater Japan Private Women's Hygienic Association. See "Ōbei ryokōchū ni kenbun seshi ichi nisetsu," *Fujin eisei zasshi* 1902, *155*: 1–14.
19. Weindling, *Health, Race, and German Politics*, p. 7.
20. German theorists mentioned in Ōsawa's 1904 work include sociologist Eduard Gumplowicz, a medical practitioner and advocate of contraceptive devices

W. Mensinga, professor of obstetrics at University of Freiburg Alfred Heger, bacteriologist who served as the director of the Department of Health in the Ministry of Welfare Martin Kirchner, and social hygienist Alfred Blaschko trained in dermatology.

21. Weindling, *Health, Race, and German Politics*, p. 9. The German notion of "social hygiene" and the French notion of "social medicine" were often used interchangeably. For social hygiene in Japan, see Nihon Kagakushi Gakkai, ed., *Nihon kagaku gijutsushi taikei*, Vol. 25, *Igaku* Part 2 (Tokyo: Dai-ichi Hōki Shuppan, 1967), pp. 73–104.
22. For instance, Ōsawa published a couple of articles regarding the impact of civilization on the physical and mental quality of humans in 1905. See Nihon Seirigaku Kyōshitsushi Henshū Iinkai, *Nihon seirigaku kyōshitsushi*, Vol. 1, p. 276. For degeneration, see Ian Dowbiggen, "Degeneration and Hereditarianism in French Mental Medicine, 1840–1890: Psychiatric Theory as Ideological Adaptation," in *The Anatomy of Madness, Essays in the History of Psychiatry*, Vol. 1, *People and Ideas*, ed. W.F. Bynum, Roy Porter, and Michael Shepherd (London: Tavistock, 1985), pp. 188–232; Daniel Pick, *Faces of Degeneration: A European Disorder, c. 1848–c. 1918* (Cambridge: Cambridge University Press, 1989), and Matsubara Yōko, "Meiji-matsu kara Taishō-ki ni okeru shakai mondai to 'iden,' " *Nihon bunka kenkyūjo kiyō* 1996, *3*: 155–169.
23. Pick, *Faces of Degeneration*, p. 11.
24. Weindling, *Health, Race, and German Politics*, p. 9.
25. Ōsawa Kenji, *Shaiteki eisei taishitsu kairyōron* (Tokyo: Kaiseikan, 1904), pp. 26–27.
26. Ibid., pp. 28, 49–50.
27. Ibid., p. 37.
28. Ibid., pp. 47, 79–85.
29. Ibid., p. 71.
30. Daniel J. Kevles, *In the Name of Eugenics: Genetics and the Uses of Human Heredity* (Berkeley: University of California Press, 1985), p. 92 and note 29 on p. 325; Weindling, *Health, Race, and German Politics*, pp. 293–294; and Atina Grossman, *Reforming Sex: The German Movement for Birth Control and Abortion Control, 1920–1950* (New York: Oxford University Press, 1995), p. 16.
31. Ōsawa, *Taishitsu kairyōron*, pp. 55–56.
32. Ibid., pp. 70–71.
33. The book, published by Ōkura Shoten in Tokyo, is 118 pages in length.
34. See Kathryn Ragsdale, "Marriage, the Newspaper Business, and the Nation-State: Ideology, in the Late Meiji Serialized *Katei Shōsetsu*," *Journal of Japanese Studies* 1998, *24* (2): 229–255. As for the *Hōchi* marriage detective service (Hōchisha Anshinjo), see its advertisement in the back of Ōsawa, *Tsūzoku kekkon shinsetsu*. For agencies investigating lineage, see Fujino, *Nihon fashizumu to yūsei shisō*, pp. 107, 392–393.
35. Ōsawa, *Tsūzoku kekkon shinsetsu*, pp. 1–2.
36. Ibid., pp. 24–25.
37. Ibid.
38. Pick, *Faces of Degeneration*, p. 92.
39. Ibid., p. 89.
40. Ōsawa, *Tsūzoku kekkon shinsetsu*, pp. 1–2. Here I translate *minzoku* as "race." However, as Kevin Doak suggests, the meaning of "*minzoku*" can only be understood within a process of discursive practice. See his "Culture, Ethnicity, and the

State in Early Twentieth-Century," in *Japan's Competing Modernities: Issues in Culture and Democracy, 1900–1930*, ed. Sharon A. Minichiellllo (Honolulu: University of Hawaii Press, 1998), pp. 181–205.

41. For the process of redefining middle class in the Meiji period and the significance of education for the new middle class people, see David R. Ambaras, "Social Knowledge, Cultural Capital, and the New Middle Class in Japan, 1895–1912," *Journal of Japanese Studies* 1998, *24* (1): 1–33.
42. Ōsawa, *Tsūzoku kekkon shinsetsu*, pp. 95 and 108.
43. Ibid., pp. 183–189.
44. As for Ōsawa's definition of physiology, see his textbook, *Seirigaku, Nihon Joshi Daigaku kōgi*, no. 9 (Tokyo: Kanda Seibidō, 1928), p. 49.
45. For Nagai's leadership of a eugenics movement, see Suzuki, *Nihon no yūseigaku*, pp. 93, 107, 144, 153–157, 167–168; and Fujino, *Nihon fashizumu to yūsei shisō*, Chapters 1, 3, 4, and 5.
46. Ōsawa, *Tsūzoku kekkon shinsetsu*, pp. 48–49.
47. Ibid., pp. 48–71. In the Meiji period, the *bushidō*, a class-specific tradition of the Tokugawa era, was often used to represent class-encompassing Japanese identity. See Unoura, "Samurai Darwinism," and Low, "The Japanese Nation in Evolution," p. 227.
48. Ōsawa, *Tsūzoku kekkon shinsetsu*, p. 4.
49. Ōtsuka Miyao, *Shinpan Meiji ishin to Doitsu shisō*, ed. Yamashita Takeshi (Tokyo: Nagasaki Shuppan, 1977), pp. 322–335; and Hashikawa Bunzō, *Kōka monogatari* (Tokyo: Chikuma Shobō, 1976).
50. The Emperor Meiji reigned Japan between 1868 and 1912 (the Meiji period). On the mobilization of medical science in the construction of racial difference, see Yuki Terazawa's contribution to this volume.
51. See Sumiko Otsubo, "Women Scientists and Gender Ideology in Japan," in *A Companion to the Anthropology of Japan*, ed. Jennifer Robertson (Oxford: Blackwell, 2005).
52. Suzuki Yūko, *Joseishi o hiraku*, Vol. 1, *Haha to onna: Hiratsuka Raichō to Ichikawa Fusae o jiku ni* (Tokyo: Miraisha, 1989), pp. 50–51; and Miyake Yoshiko, "Kindai Nihon joseishi no saisōzō no tame ni: Tekisuto no yomikae," Kanagawa Daigaku Hyōron Henshū Senmon Iinkai, ed., *Kanagaka Daigaku hyōron*, Vol. 4, *Shakai no hakken* (Tokyo: Ochanomizu Shobō, 1994), p. 65.
53. Ōsawa, *Tsūzoku kekkon shinsetsu*, pp. 555–564.
54. Ibid., p. 561.
55. Daniel Pick has discussed the increasing use of medical metaphors in describing a nation's historical and social phenomenon in late-nineteenth-century Europe. See his *Faces of Degeneration*, pp. 97–99.
56. Ōsawa, *Tsūzoku kekkon shinsetsu*, p. 564. Ōsawa seems to have been influenced by Katō Hiroyuki's theory to equate a society to an organism (*shakai yūkitaisetsu*) and glorification of the altuistic *bushidō*. See Unoura, "Samurai Darwinism."
57. Sumiko Otsubo, "Engendering Eugenics: Feminists and Marriage Restriction Legislation in the 1920s," in *Gendering Modern Japanese History*, ed. Barbara Molony and Kathleen Uno (Cambridge, Mass.: Harvard University Asia Center, forthcoming).
58. Ōsawa, *Taishitsu kairyōron*, p. 84; and *Tsūzoku kekkon shinsetsu*, p. 557.
59. Ōsawa Kenji's article was originally published in the journal *Sei* (November 1920). It was reprinted as "Karyūbyō danshi kekkon seigen-hō hiketsu no fujōri," *Josei dōmei* 1920, *3*: 47–48.

60. Ōsawa Kenji, "Shugai to kinshu hōhō," *Kuni no hikari* 1907, *167*: 20–22.
61. Katō Junji, *Nemoto Shō-den: Miseinensha inshu kinshu-hō o tsukutta hito* (Nagano: Ginga Shobō, 1995), pp. 167–171.
62. "Dai nijūrokkai Teikoku Gikai Kizokuin gijiroku kiroku bassui Ōsawa igaku hakushi no shugai dai enzetsu," *Kuni no hikari* 1910, *202*: 18–24.
63. See, e.g., Chalmers Johnson, *MITI and the Japanese Miracle: The Growth of Industrial Policy, 1925–1975* (Stanford: Stanford University Press, 1982). See also the assessment of this influential view in Morris-Suzuki, *The Technological Transformation of Japan: From the Seventeenth to the Twenty-first Century* (New York: Cambridge University Press, 1994), pp. 72–77; and Andrew Gordon, *A Modern History of Japan: From Tokugawa Times to the Present* (New York: Oxford University Press, 2003), p. xii.
64. See e.g., Fujino, *Nihon fashizumu to yūsei shisō*, p. 40.
65. See Fujime Yuki, *Sei no rekishigaku: Kōshō seido, dataizai taisei kara Baishun Bōshi-hō, Yūsei Hogo-hō taisei e* (Tokyo: Fuji Shuppan, 1997), pp. 320–321; Kondō Kazuko, "Onna to sensō" in *Onna to otoko no jikū: Nihon joseishi saikō*, Vol. 6, ed., Okuda Akiko, *Semegiau onna to otoko: Kindai* (Tokyo: Fujiwara Shoten, 1995), p. 481.
66. Sheldon Garon has analyzed complex state–society relations, see his "Rethinking Modernization and Modernity in Japanese History: A Focus on State-Society Relations," *Journal of Asian Studies* 1994, *53* (2): 346–366.
67. "Minzoku Yūsei Hogo hō-an iinkai giroku (sokki) daini-kai," in *Dai nanajūy-onkai, Teikoku Gikai Shūgiin iinkai giroku, Shōwa 13, 14 nen, in Teikoku Gikai, Shūgiin iinkai giroku*, microfilm ed., reel 31 (1938–1939; reprint, Rinsen Shoin, 1990), p. 369.
68. See Suzuki Zenji, "Yūzenikkusu ni taisuru Nihon no han'nō," *Kagakushi kenkyū* 1968, *87*: 129–136, on p. 135.
69. Andrew Gordon states that "a profound anxiety that something was being lost in the headlong rush to a Western-focused modernity surfaced with increasing intensity in the 1880s and 1890s. This worry pushed intellectuals to improvise new concepts of Japanese 'traditions.' It also linked up with the fear of social disorder and political challenge among state officials. They responded by putting in place oppressive limits on individual thought and behavior." See his *A Modern History of Japan*, p. 94. For invented traditions in Japan, see Stephen Vlastos, ed., *Mirror of Modernity: Invented Traditions of Modern Japan* (Berkeley: University of California Press, 1998).
70. Ōsawa, *Taishitsu kairyōron*, p. 37; and *Tsūzoku kekkon shinsetsu*, pp. 93–95, 195–196.

4

Racializing Bodies through Science in Meiji Japan: The Rise of Race-Based Research in Gynecology

Yuki Terazawa

Introduction

In his 1908 paper discussing the reproductively active years of women of various ethnic backgrounds, physician Yamazaki Masashige (1872–1950) emphasizes the idea that many different "races" reside in the Japanese empire other than the Japanese race, which he describes as "the race descended from the imperial line" (*tenson shuzoku*).[1] These so-called inferior races included the Ainu, the Chinese in Taiwan, Taiwanese aborigines, and the people who inhabited the Ryūkyū islands (the Ryūkyūans). Discussing the relations between the Japanese and these other races, Yamazaki draws on Social Darwinist thinking: "According to the law in which the superior conquers the inferior, weaker races will be subordinated by stronger ones. These [inferior] races would either assimilate to a superior one or perish. [As such,] they will never preserve the original racial characteristics."[2] Believing that these non-Japanese "races" would eventually become extinct, Yamazaki felt it urgent to study their racial traits, including differences among the different races in the onset of menstruation and menopause, while these racial groups still existed. Yamazaki was one of numerous Meiji scientists who appropriated from Europe and the United States the notion of race as a scientifically valid category along with Social Darwinist ideas. Focusing on Yamazaki's paper, I examine the way sexed and racialized bodies emerged from scientific and medical discourses in Japanese history. I also explore how scientific and medical discourses on race were developed in conjunction with discourses and policies associated with Japan's nation- and empire-building projects in the late nineteenth and early twentieth centuries. This case study shows that scientific and medical research, while at times maintaining a certain autonomy, was never immune to political, social, and economic forces.

DEBATES ON "RACE" IN MEIJI JAPAN

Race was an important concept in European scientific and medical practices during the eighteenth and nineteenth centuries. Attempts to classify people into different groups based on the racial characteristics manifested in human bodies constituted a significant part of medical and anthropological research. Scholars used increasingly sophisticated methods and instruments, including photography, for measuring various body parts and identifying racial traits with precision. In the late nineteenth and early twentieth centuries, many anthropologists used the evolutionary paradigms made available by Social Darwinism to divide people into different racial groups. Differences in specific physical characteristics, such as the size of the skull, were used not only as markers for classifying people but also as a means of locating certain racial groups within a linear civilizing process that mankind inevitably is meant to undergo. By replacing "species" with "race" in the Darwinian struggle for survival, they also asserted that some races were destined to perish while others would prosper.[3]

As a scientific concept, race was introduced and popularized in Japan in the late 1870s and in the 1880s through the adoption of Social Darwinism, eugenics, and anthropological methods. Scientific studies of race in Japan had been initiated by European and American scholars who went to Japan beginning in the early 1870s.[4] The American zoologist Edward S. Morse (1838–1925) introduced anthropological and biological methods in Japan for the first time in the late 1870s.[5] As a visiting professor at Tokyo Imperial University, Morse introduced Darwinian evolutionary theory, even preceding the publication of Japanese translations of Darwin's works.[6] Morse also contributed to the development of anthropology in Japan through his famous discovery and excavation of the shell mounds of Omori. Morse's interest in Japanese prehistory also led him to explore the racial formation of the people who lived in Japan during that period.

By the mid-1880s, Japanese intellectuals were engaged in vociferous discussions on the history and contemporary issues concerning the Japanese race and its relationship with other racial groups; Social Darwinist thinking from Europe and the United States played a large role in these discussions. Once Morse had introduced Charles Darwin and Herbert Spencer's evolutionary theories, prominent Japanese scholars at Tokyo Imperial University, such as Toyama Shōichi (1848–1900), Katō Hiroyuki (1836–1916), and Oka Asajirō (1868–1944), enthusiastically embraced Social Darwinism to explain the state, politics, human society and history. A version of Social Darwinism that many Japanese intellectuals adopted postulated the state as a natural organism and people as individual cells. Based on this thinking, these scholars argued for the importance of protecting the interests of the state, which presumably constituted the core of this living organism, even when it meant sacrificing the well-being of individuals. Furthermore, they used Social Darwinism to justify economic, political, and social inequality among individuals as a natural outcome of the theory of natural selection with its need for continuous struggle to ensure the ongoing improvement of the race. Japanese thinkers also extended the notion of individuals competing with each other in a "struggle for existence" to nations and racial groups, which they envisioned as going through a similar process.[7]

Accepting the notion of an evolutionary scale indicating the level of advancement reached by each nation, Japanese intellectuals from the Meiji period generally believed in the inferiority of Japanese people vis-à-vis European populations. However, this did not lead them to argue that the Japanese were destined to be defeated in the competition between nations and racial groups. Rather, they suggested that by implementing adequate social, economic, educational and public health policies, Japan would be able to improve its citizen's physical and mental capacity to advance its civilization, and to compete against European nations and the United States.

This thinking was demonstrated in the debates about whether Japan should abolish restrictions on the areas where foreigners were allowed to reside within Japan and give them freedom to choose their own residences. Those who supported mixed residency argued that the presence of Westerners would promote economic, entrepreneurial, and cultural developments in Japan. Some even proposed that the Japanese should promote interracial marriages with Westerners in order to strengthen their racial stock. Others vehemently opposed mixed residency because they thought it would result in Westerners taking advantage of them economically and monopolizing Japanese resources. Drawing on Herbert Spencer, some of them suggested that interracial marriages between the Japanese and Westerners would lead to the demise of the Japanese race because of the rule that the blood of the superior race would subordinate that of the inferior race when they were blended by marriage.[8] Although there were a number of viewpoints in this debate, they were all framed by Social Darwinist thinking.

In addition to contemplating relations between the Japanese and European races, both European and Japanese scholars sought to redefine various racial and ethnic groups in East Asia by using newly introduced anthropological methods. In addition to Edward Morse, German scholars and physicians teaching at Tokyo Imperial University led these research efforts. For example, based on the data obtained by measuring skeletal specimens, anatomist Wilhelm Dönitz presented a theory about the racial formation of modern Japanese people. He hypothesized that the Japanese race derived from the mixing of two different races: the Malay and the Mongoloid.[9] Dönitz claimed that the Mongoloid race included two different types, one of which was the Ainu.

Contrary to Dönitz, who formed his hypotheses based almost exclusively on people's physical traits, Erwin von Baelz (1849–1913), a professor of internal medicine at Tokyo Imperial University, incorporated differences in cultural customs into his racial typology. He regarded the Ainu as a separate racial group from the Japanese belonging to the Caucasian race. He divided the Japanese group into two distinct types, both of which derived from the Mongoloid race. The first was what he called the Chōshū type, a group whose ancestors migrated from the Chinese continent through Korea and spread through the Chōshū area—the southwestern tip of Japan's main island. They possessed a slender body, a long head, a long face, up-turned eyes, a nose of a medium height, and a small mouth. Baelz claimed that this type was often found among upper-class Japanese as well as upper-class Chinese and Koreans. A second type called the Satsuma type also belonged to the Mongoloid race but resembled the Malays. According to Baelz, a larger number of Japanese people, particularly commoners, belonged to this group. Their ancestors also migrated from the Korean peninsula, but unlike the first group, they initially settled in southern Kyūshū—one of the four Japanese islands located in the south—before they conquered the rest of Japan. Their facial and bodily traits were marked by short and stocky stature, short head, wide and short face, high cheekbones, eyes that were less slanted, a flat nose, and a large mouth.[10]

The Korobokkuru Debate and the "Original" Japanese

Following the lead of these European scholars, Japanese anthropologists also embarked on research on races in Japan and its vicinity. During the 1880s, while Social Darwinism became increasingly popular among Japanese intellectuals, they began to show a strong interest in studying the racial identity of the Japanese, especially that of the "original inhabitants" on Japanese islands during prehistoric ages. This interest culminated in the so-called Korobokkuru controversy. This debate centered on the question of who lived in the Japanese islands during the Stone Age, before various groups of people migrated from the Eurasian continent and islands in Southeast Asia and the Pacific. The leading anthropologist Tsuboi Shōgorō (1863–1913) and his followers maintained that the "original Japanese" were the so-called Korobokkuru tribes who appeared in Ainu mythology, and who were presumably forced out by the thriving Ainu people at the time. Another group of anthropologists, including the prominent physical anthropologist and anatomist Koganei Yoshikiyo (1859–1944), argued that the legendary Korobokkuru people were in fact an Ainu tribe that inhabited Japan during the Stone Age.[11]

Tsuboi's 1888 research trip to Hokkaidō convinced him that his own hypothesis was correct. After excavating shell mounds and other prehistoric remains in Hokkaidō, Tsuboi asserted that the Stone Age people (i.e., the Korobokkuru people) possessed specific customs and cultural artifacts, such as living in pits and making clay pottery and stoneware, which were different from the Ainu culture.[12] Taking issue with Tsuboi was Koganei, who accompanied Tsuboi on the same research trip. Koganei developed his own theory based on reports that some Ainu tribes in islands north of Hokkaidō had engaged in the same cultural practices ascribed to Stone Age people in question. Tsuboi's thesis eventually lost credibility due to the findings of Torii Ryūzō (1870–1953) on his 1899 research trip to the Chishima islands.

Torii discovered that the Ainu tribes in the Chishima islands lived in pits and used similar stoneware and clay pottery. The Ainu people whom he interviewed demonstrated that such customs had been handed down to them by their ancestors, not left by other peoples. Moreover, they did not have any legends about aborigines who had lived on the land before they settled there.[13] Such facts suggested the probability that the mythological Korobokkuru people were an offshoot of the Ainu tribes.

What is of interest here is not the validity of these various arguments, but the preoccupation that Japanese anthropologists and the general public developed in the Ainu as a racial "other." Both Tsuboi's and Koganei's factions shared the basic understanding of the Ainu as an inferior, uncivilized, and "dying" race.[14] Such attitudes about a "primitive race" were only made possible by the Japanese intellectuals' appropriation of ethnocentric interest, methods, and attitudes as embedded in racial theories produced in Europe and the United States.

The ways in which Japanese scholars discussed racial differences involving racial or ethnic groups in East Asia were more complex than similar debates in Europe. As opposed to Europeans, who could often posit an unambiguous boundary between themselves and "non-European" races, Japanese scholars could not deny certain affinities between the Japanese and what they considered other racial groups in East and Southeast Asia. Japanese intellectuals often strategically cited differences or affinities between the Japanese and other races in East Asia to pursue political agendas. In order to clarify how, when, and why specific strategies of exclusion and inclusion were adopted, we need to conduct extensive research encompassing diverse fields and historical periods.[15] The following case study aims to contribute to such scholarship.

Yamazaki Masashige and Women's Bodies

Many physical anthropological studies from the late nineteenth century focused on studying the differences between the Ainu and the Japanese, racial categories that many anthropologists had accepted as indisputable.[16] However, by the turn of the twentieth century, when Ryūkyū and Taiwan had become Japanese colonies, some Japanese anthropologists and physicians asserted that the racial composition of people living within the Japanese empire was more complex than a simplistic division between the Ainu and the Japanese. For example, the obstetrician–gynecologist Ogata Masakiyo (1864–1919) severely criticized German researchers for failing to classify the Japanese and groups such as the Chinese, the Koreans, and the Ainu as separate races.[17] Some Japanese anthropologists began publishing papers on the anatomical characteristics of Koreans and a Japanese outcast group that had been called the "*eta*" or "*kawata*" during the Tokugawa period (1603–1868). Thus, Japanese researchers developed a great deal of interest in clarifying racial divisions, including the physical and mental traits specific to each race, in Japan and nearby countries. Yamazaki Masashige's (1872–1950) paper "On Menstruation of Women of Four Races: The Japanese, the Ainu, The Ryūkyūans, and the Chinese" responded to the kind of criticisms leveled by Ogata and attempted to establish a more complex framework for dealing with racial differences among East Asian peoples.[18]

Yamazaki was a leading obstetrician–gynecologist during this period. He followed a typical career trajectory for an elite physician, studying at the medical school of the Tokyo Imperial University and receiving his graduate education in Germany. Upon completing his medical studies, Yamazaki assumed supervisory positions at several different publicly funded hospitals. The above-mentioned paper was written while he was presiding over the gynecology and obstetrics division of Kumamoto hospital, an institutional affiliation that facilitated access to his research material—women's bodies. This work was also facilitated by the professional network he developed while attending Tokyo Imperial University and working at public hospitals, both of which provided opportunities to collect data on women of ethnic minorities.

Yamazaki's paper begins by underlining the importance of studying women's reproductive capacity for national purposes and deploring the fact that Japan lagged far behind European nations and the United States in this area of research. Studies by European and American researchers, he observes, indicate that the timing of menarche and the onset of menopause differ in response to a variety of environmental factors, including geographic location, climate, custom, the degree of civilization, status, profession, living standards, nutrition, and health. European researchers had also discovered that the length of active reproductive years depended on specific conditions pertaining to each individual, such as certain hereditary traits, personality, and physique. Yamazaki contends that, even though some Japanese physicians had conducted statistical research similar to that in Europe and the United States, they had not sufficiently explored racial differences pertaining to women living in various parts of the Japanese empire.[19]

Filling this gap was in fact Yamazaki's main goal. His study is based on Social Darwinist notions, appropriated from European studies in which physicians attempted to correlate the reproductive physiology of different races and classes to the degree of cultural and material progress each group had supposedly attained. While Yamazaki shared the basic Social Darwinist assumptions of European physicians, however, his justification for claiming the Japanese race to be the superior to other races in the Japanese empire was unique. His science is mingled with ideas about the mythical past of a mighty Japanese race that had presumably driven away inferior races such as the Ainu from Japan's main island.

Yamazaki and Japan's Colonial Project

Yamazaki attributes the dispersion and assimilation of the Ainu to the spread of the cultural and political influence of the Yamato race, or as he calls it, "the race that descended from the Japanese imperial family (*tenson shuzoku*)."[20] During this conflation of mythical, prehistoric, and historic events, Yamazaki even invokes the court-appointed general Sakanoueno Tamuramaro's (758–811) campaign to fight against the *ezo*—the "toad barbarians" in the East—in the late eighth and early ninth centuries as evidence of the strength of the Japanese race and the biological feebleness of the Ainu.[21]

Yamazaki pursues this argument by describing the dire health conditions of Ainu communities and the rapidly declining Ainu population. He contends that the Ainu had once been a physically robust people, but had deteriorated because of the conversion of their hunting fields to agricultural settlements, depriving them of sufficient game to maintain their traditional meat-based diet.The rate of infant mortality, according to Yamazaki, was also very high among the Ainu.[22] Yamazaki ignores, however, the Meiji government's aggressive colonial policies, which had developed agricultural settlements in Hokkaidō and thereby deprived the Ainu of their land. Another reason for the decline of the Ainu population was new diseases transmitted by Japanese colonizers.[23] In Yamazaki's view, however, inferior races such as the Ainu were fated to vanish as a result of natural law (*shizen no tensoku*).[24] In this way, the net effect of Yamazaki's Social Darwinist argument is to conceal the political and economic processes that were transforming the lives of the Ainu.

In contrast to his discussion of the Ainu, Yamazaki emphasizes the racial affinity between the Ryūkyūans and the Japanese, leading him to conclude that as time passed the former would be increasingly assimilated to the latter, finally losing their particular racial characteristics. Although he dismisses Japanese colonial policies that subjected Ryūkyūans to Japanese rule, Yamazaki makes the curious argument that Ryūkyū had been annexed to Japan earlier in the Meiji period than any of the other colonies because of the Ryūkyūans' racial affinity to the Japanese. In this view colonization is a natural process caused by the presumed biological and cultural similarity of the two races. As such, the Ryūkyūans' assimilation to the Japanese was an inevitable result of biological principles. According to Yamazaki, this process of acculturation began immediately after the annexation and had been rapidly advancing through interactions with the Japanese and their superior civilization.[25]

Yamazaki was also certain that the Chinese people and Taiwanese tribes in Taiwan would eventually be assimilated into the Japanese race—either that or they would vanish. In Yamazaki's view these races had benefited by adopting Japanese ways after being subjected to Japanese rule following the 1895 Sino-Japanese War. Unfortunately, there were some Taiwanese aborigines—Yamazaki refers to them as "raw" aborigines (*seiban*)—who, unlike "mature" aborigines (*jukuban*), had refused to receive the benefits of civilization. Yamazaki writes that these tribes resided deep in the mountains and maintained their barbaric customs, refusing to adopt even Chinese civilization. Yamazaki predicted that these "raw" natives would follow the same fate as the Ainu—losing their indigenous customs and eventually disappearing as a race.[26]

The predicted permanent loss of original racial traits in these presumably inferior races, whether by assimilation or extinction, posed a serious challenge to Yamazaki, who considered it extremely important for scientists to study such distinct racial characteristics before they disappeared.[27] This sense of urgency was shared by many other Japanese scholars and intellectuals in diverse fields. Looking upon these "dying races" as objects of intellectual inquiry, the Japanese researchers deemed it imperative to preserve their languages, customs, and artifacts in the form of scholarly research, collections of indigenous literature, and museum exhibits.[28] While some Japanese scholars found their way into indigenous communities as researchers, these intrusions almost always occurred in collaboration with other types of colonial projects in such diverse fields as politics, business and education.

Yamazaki's research, too, was facilitated by the network of colonial institutions, particularly the publicly funded medical schools and hospitals presided over by the colonial administration. Indeed, it was by contacting physicians and educators working on educational projects and in public medical facilities in the Ryūkyū islands, Taiwan, and Ainu communities in Hokkaidō that he obtained most of his research data. For example, Yamazaki collected data on Ryūkyūan women with the help of assistant directors of the Okinawa prefectural hospital. Likewise, the physician Takagi Eisen, who directed research on Chinese women, was one of Yamazaki's friends who had once practiced in Kumamoto and was working at that time as a government-appointed physician in southern Taiwan.[29] Data collection on Ainu women was carried out by Oyabe Zen'ichirō (1867–1941), an American-educated missionary teacher who was stationed in the Iburi area in southwestern Hokkaidō. Oyabe's own research was supplemented by information gathered by Japanese educators who resided in other towns in the area.[30]

COLLECTING DATA

In order to gather data, Yamazaki and his collaborators needed the cooperation of women who were able and willing to provide information about their menstrual cycles. In this regard, Yamazaki confronted many difficulties in obtaining data from women of minority ethnicities, who were resistant to discussing menstrual issues with male researchers.[31] Even among willing subjects he encountered other problems. Many of the women, for instance, did not know their correct date of birth or the date of their first menstruation. Yamazaki was condescending to women of ethnic origins about their ignorance of their bodily processes and their unwillingness to share information about their menstruation cycles with the researchers—an attitude he viewed as an indicator of cultural backwardness. For example, Yamazaki remarks that the type of research he wanted to conduct was extremely difficult to carry out in Taiwan because Chinese women were in the habit of keeping matters of menstruation strictly among women. According to Yamazaki, this practice was a manifestation of their "obstinate adherence to old customs" dictating that contact with men was distasteful.[32]

In their research with the Ainu, Yamazaki and his colleague Oyabe encountered a general aversion to interacting with the Japanese. They also found that both Ainu women and men felt very ashamed when asked to talk about their private parts (*inbu ni kansurukoto*), including menstruation. These researchers considered the Ainu's ignorance about menstruation appalling. Yamazaki lists his findings as follows: Ainu mothers taught their daughters about menstruation, but their explanation consisted only of several words; some Ainu women never even learned the Ainu words for menstruation; and the men, he believed, often knew nothing about menstruation since Ainu women never talked with them about menstruation, even with their fathers and husbands. Yamazaki and Oyabe also refer to the Ainu's association of menstruation with defilement, a belief that prevented women from worshipping gods during their menstrual periods, as evidence of backward attitudes and beliefs.[33] The difficulties of conducting research on women of ethnic origin provided Yamazaki an excuse for relying on a small number of samples, and allowing him to declare that the data that he managed to collect were invaluable despite their modest scope.[34]

Yamazaki's frustrations with collecting data from women of ethnic origin sheds light on the process by which women were transformed into modern subjects who could provide biographical and physiological information about themselves in a language intelligible to medical researchers. Yamazaki wanted the women to be compliant informants, but transforming them into model interviewees required an exhaustive colonizing process. Women had to be taught to communicate in the proper way, whether it was in their native language or the language of the researchers, and it was necessary to equip them with new ideas and attitudes in order to break down their deep-seated reluctance to discuss reproductive issues with strangers and men in general.

This educational process involved replacing an existing local understanding about bodily phenomena with one provided by modern medical science. In order for this to happen, women had to recognize the authority of medical researchers in a form that would make them responsive to the researchers' requests. In other words, the women had to be made into acquiescent modern subjects with certain views and attitudes who would collaborate with the modern medical establishment in accumulating discursive knowledge. The efforts by Yamazaki and his colleagues to gather knowledge about indigenous women's bodies was thus facilitated, and in some cases enabled, by state-supported medical projects that were first established in Japan and gradually extended to its colonies.

Yamazaki's accounts of his failure in conducting research on Taiwanese aborigines, however, indicate the limits of colonial institutional practices beyond mainland Japan in 1908, the year he published his study. Japanese colonial power in its military, political, and cultural forms had not yet penetrated into the society of these Taiwanese tribes, a fact that made it difficult or impossible for Yamazaki to gather data from them. Yamazaki believed that acculturating these so-called savages would be extremely difficult; he depicted them as barbaric, violent, cruel beings who preferred killing to civilized means of resolving disputes.[35] However, the militant customs he discussed with such hostility, frustration, and fear also effectively prevented

Table 4.1 Average age of menarche for Japanese, Ainu, Ryūkyūan, and Chinese women

	Japanese	Ainu	Ryūkyūan	Chinese
Number of informants	23,754	80	184	135
Average age of the onset of menstruation	15 years, 1st month	15 years, 2nd month	16 years, 1st month	16 years, 7th month

Source: Yamazaki Masashige, "Nihon, Ainu, Ryūkyū, oyobi Shina yon shuzoku fujin no gekkei ni tsuite," *Ogata fujin kagaku kiyō* 1908, 2: 148–170.

Japanese colonial power from reaching these aboriginal Taiwanese tribes and turning their women into docile subjects ready to collaborate with modern medical research.

The difficulties in obtaining data from women of ethnic origins were reflected in the immense discrepancy between the number of Japanese informants and those of women in other categories. In order to calculate the average age of menarche for Japanese women, Yamazaki used research results of thirteen other scholars as well as his own: the total number of women interviewed was 23,754. In comparison the number of informants from ethnic communities was significantly smaller; 80 Ainu women, 184 Ryūkyūan women, and 135 Chinese women[36] (see table 4.1). This data suggested that Japanese women began menstruating at the earliest age, 15 years and 1 month, followed by the Ainu women, whose average age for menarche was 15 years and 2 months. The average ages of the Ryūkyūan and Chinese women were 16 years and 1 month, and 16 years and 7 months respectively.[37]

Understanding the Data

Having obtained these results, Yamazaki proposes an initial hypothesis in his paper: that climate is the major factor determining the average age of menarche. The warmer the climate, the earlier women would start menstruating. According to this theory, Chinese women in Taiwan would commence menstruation the earliest, followed by the Ryūkyūans, the Japanese, and the Ainu.[38] However, Yamazaki's data obviously contradict this assumption. In order to explain this paradox, Yamazaki asserts that the climatic variations among Hokkaidō, the Japanese islands, the Ryūkyūan islands, and Taiwan were not as significant as the Japanese imagined.[39] If people considered Hokkaidō's altitude, he argues, they would realize that its position was actually comparable to other "civilized nations" in Europe.[40] Yamazaki supports his claim by citing the German physician Erwin von Baelz, who suggested that the climate of Hokkaidō was similar to that of his home country.[41] At the same time, Yamazaki alleges that although some parts of Taiwan belong to semi-tropical zones, the heat of the summer is mitigated to a large extent because it is an island surrounded by the ocean, thus making its climate comparable to that of Kyūshū.[42] In this way, Yamazaki portrays the climatic influences as minor, if not completely irrelevant.

If climatic variation did not explain the differences in the timing of menarche among women of diverse racial groups, what did? At this point Yamazaki invokes the notion of cultural practices. Viewed through his ethnocentric lens, these practices provide an explanation for differences in race-specific female reproductive physiology. According to Yamazaki and his collaborator Takagi, Chinese women in Taiwan commenced menstruation later than women of other racial groups because their adherence to backward customs prohibited them from receiving both the "social stimulus" (*shakaiteki shigeki*) and physical exercise these researchers deemed indispensable for developing a healthy body. Takagi observes that Chinese women, particularly those from upper-class families, lived in the dark interior of their mansions and never interacted with men. Some of them, he continues, even used the lavatory inside their rooms. Yamazaki concludes that such a sedentary and withdrawn lifestyle, reinforced by the practice of foot-binding, results in a weak constitution. Chinese women, in his view, also lacked access to "social stimuli" due to the presumed fact that their society lagged behind the Japanese in terms of worldly progress and civilization.[43]

Yamazaki implies that Ryūkyūan women shared some of the backward customs maintained by Chinese women in Taiwan; however, the degree to which they had adopted modern civilization was greater than their Chinese counterparts.[44] Because of this, he suggests that Ryūkyūan women generally started menstruation earlier than Chinese women. While cultural backwardness and a lack of physical exercise explained the relatively late ages at which Chinese and Ryūkyūan women started menstruation, Yamazaki focuses solely on the benefits of exercise for rationalizing Ainu women's early menarche. Unlike the inactive and secluded life of Chinese women, Yamazaki describes the majority of Ainu women as engaging in fishing and farming in ways not so different from men. According to Yamazaki, this helped Ainu women to develop a stronger constitution.[45]

While this reasoning solved the problem of why Ainu women experienced menarche at an earlier age than Chinese and Ryūkyūan women, it does not explain why some Ainu women started menstruation earlier than Japanese women. Since Ainu society was presumably so culturally backward (*kaimei no teido otori*) and Ainu people lived in a colder climate, these research results presented him with a disturbing problem.[46] In response, Yamazaki develops the idea of "innate racial characteristics" embodied in the body's physiology.[47]

Yamazaki believed that these "racial peculiarities" would be mitigated and even offset by climatic and other factors over a long period; however, such innate racial traits would sometimes become a major determinant of certain physiological phenomena. Yamazaki cites the example of English women born in India, who started menstruating later than Indian women, just as women in England did. In this case, intrinsic, racially specific physiological processes overpowered the climatic influence.[48] Yamazaki suggests that there must be inborn racial particularities governing the body's physiological processes. These were responsible for the Ainu women's early menstruation, even though modern medical science had not yet elucidated the physiological mechanism for these mysterious race-based attributes.[49]

The notion of "racial peculiarities" served as a convenient *deus ex machina* that could be used to explain away contradictory evidence as a product of unspecified racial differences. The selective application of racially determined bodily differences sustained Yamazaki's racial hierarchy, along with its underlying Social Darwinist assumptions, by preventing a rethinking of the theoretical framework, despite the presence of discordant data. Perhaps an even more important consequence was that the concept of race-based biological difference, in collaboration with other scientific ideas and practices, substantiated and legitimized "race" as a category endowed with scientific authority. This is especially true if one considers that Yamazaki's proposal of race-specific differences as the cause of differences in the timing of menarche was little more than an unsubstantiated assumption. In Yamazaki's thinking, however, this notion played a major role in reifying racial differences and for sustaining the Social Darwinist theories his research data supposedly support.

Yamazaki also invokes Social Darwinism to explain the reproductive cycles of women from the same racial group but different social backgrounds. However, he only applies this analysis to Japanese women, not other ethnic minorities. In fact, he fails to mention any diversity at all among women of ethnic communities. By consigning the women of each ethnic group to a singular category, Yamazaki reinforces the idea that their bodies were characterized by overriding "racial" traits. There are also striking methodological problems with his investigation of the influence of class, occupation, geography, and other factors on Japanese women's reproductive years. In general, these analyses lack solid numerical evidence. Nor does he provide a convincing explanation of exactly how Social Darwinism accounts for the supposed differences among different classes within the same racial group.

His ill-defined research method is illustrated by the way Yamazaki categorized 1,583 Japanese female informants according to the routes through which he gained access to them as research samples (see table 4.2). The first group included 900 female patients who visited the obstetrics and gynecology division of the Kumamoto prefectural hospital. Their average age of menarche

Table 4.2 Average age of menarche for Japanese women of different categories at the Kumamoto Prefectural Hospital

	Regular patients	Students in nursing and midwifery	Licensed prostitutes[a]
Number of informants	900	112	572
Average age of menarche	14 years, 10th month, 15th day	14 years, 6th month, 12th day	15 years, 1st month, 11th day

Note: [a] While Yamazaki does not specify in his paper, they most likely visited the hospital for state-mandated examinations and treatment of venereal diseases.

Source: Yamazaki, p. 126.

was the fifteenth day of the tenth month of the year when they were 14 years old. The second category was 112 students of nursing and midwifery, who on average began menstruating on the twelfth day of the sixth month at the age of 14. The third group was comprised of 572 licensed prostitutes, who on average began menstruating the eleventh day of the first month at the age of 15.[50] Despite his presentation of this numerical evidence, Yamazaki provided no explanation for these data.

Yamazaki also classified women according to the occupation of their fathers or families. He claims that women whose families ran restaurants and hotels started menstruation the earliest, followed by daughters of fishermen, public servants, physicians, attorneys, and teachers. Next were women whose fathers were unemployed, who worked for commercial and industrial establishments, and who engaged in farming. Daughters of laborers commenced menstruation the latest of all.[51] Here Yamazaki even fails to provide numerical data or explanations for the research results.

This rudimentary presentation of the influence of social background on women's reproductive physiology is followed by very general remarks about the effects produced by class differences and the urban or rural environment in which women were brought up. Yamazaki concludes that women from "higher society" (*jōtō shakai*) tend to commence menstruation at an earlier age than those from "lower society" (*katō shakai*), a conclusion reinforced by his comment that the average age of the first menstruation of women from wealthy households was earlier than for women from poor families. In addition, he claims that women who lived in cities and towns generally began menstruating earlier than those who lived in rural areas.[52] Although the exact reasons for these differences are unclear, this section continues to show the influence of Social Darwinism in its assumption that menarche is hastened by exposure to a "civilized" lifestyle, and the rather strange corollary that women who start menstruating earlier are somehow more "advanced" than women who start later.

Taken in sum, Yamazaki's analysis reveals the emergence of medical discourses predicated on the notion of biological differences among the bodies of different classes; however, the extent to which Yamazaki explores that line of examination is quite limited. Unlike some European scholars of the time, Yamazaki does not rigorously argue that there are differences in reproductive physiology between women from the upper and middle classes and those from working class and impoverished peasant families. Moreover, he does not explicitly invoke Social Darwinist theory to explain the differences in the timing of menarche for women from different social backgrounds. Yamazaki does not seem to be overtly influenced by scientific, medical, and popular discourses of the time that were increasingly defining the minds and bodies of lower-class people, criminals, and prostitutes as deviant from those of "normal" people of upper- and middle-class backgrounds. Nor does he allude to ethnocentric discourses prevalent at that time, which described the "peculiar" living conditions and cultural habits of lower-class Japanese women and poor peasant women in rural areas. This could be partly due to the fact that at the turn of the twentieth

lifestyle had a distinct cultural pattern from that of the Ainu. Kudō, *Kenkyūshi*, pp. 83–92; Terada, *Nihon no jinruigaku*, pp. 55–59.

12. Ibid., pp. 93–96; Ibid., pp. 59–62.
13. Ibid., pp. 116, 120–124; Ibid., pp. 63, 81–82.
14. For example, Koganei believed that the Ainu race, which had once been the inhabitants of Japan, was expelled from Japan's main island due to the invasion of the presumably stronger and superior Japanese race. As a result, they were living only in Hokkaidō by the time Koganei conducted his research. More specifically, he considered the Ainu as a declining racial group compared to the thriving Japanese race which he thought possessed a more advanced civilization than the one that the Ainu maintained. Kudō Masaki, "Ikakei jinruigaku no seiritsu to sono tokushitsu," *Tōhoku rekishi kan kenkyū kiyō* 1978, *4*: 4–5. For various representations of the Ainu as a "dying race," see Richard Siddle, *Race, Resistance and the Ainu of Japan* (London and New York: Routledge, 1996), pp. 76–112.
15. In *Tan'itsu minzoku shinwa no kigen*, Oguma Eiji makes a valuable contribution on this issue. Oguma (1995). See also his more recent work, *Nihonjin no kyōkai: Okinawa, Ainu, Taiwan, Chōsen, shokuminchi shihai kara fukki undō made* (Tokyo: Shinyōsha, 1998). A discussion of Japanese appropriations of race theory and evolutionism in the larger context of the colonization of Asia and the Pacific is found in Christine Dureau and Morris Low, "The Politics of Knowledge: Science, Race and Evolution in Asia and the Pacific," *History and Anthropology* 1999, *11* (2–3): 131–156.
16. See, e.g., Y. Koganei and G. Osawa, "Das Becken der Aino und der Japaner," *MittheilungenI Aus Der Medicinischen Facultat Der Kaiserlich-Japanischen Universitat Zu Tokio*, IV. Band., 1900. A Japanese translation of this paper is found in Ogata, pp. 1227–1269.
17. Ibid., pp. 1090–1091.
18. Yamazaki, "Nihon, Ainu, Ryūkyū, oyobi Shina yon shuzoku fujin no gekkei ni tsuite," op.cit., pp. 108–177. According to Ogata's *Nihon sanka gakushi*, this paper was also published in 1909 in German as M. Yamazaki, "Ueber den Beginn der Menstruation ei den Japanerinnen, mit einem Anhang ueber die Menge bei den Chinesinnen, den Riukiu-und Ainofrauen in Japan." Ogata does not indicate where the German version was published. Ogata, p. 1762.
19. Ibid., pp. 108–110.
20. Ibid., p. 110. The Japanese word, "*minzoku*," which was used by Yamazaki for the translation of "race" as a scientific term, also connotes "people" or "ethnic group." It seems that Yamazaki took advantage of the word's ambiguity to conflate the scientific definition of "race" with "people," implying a grouping by historical and cultural factors.
21. Ibid., pp. 111–113. Whether the *ezo* from the eighth and ninth centuries was identical to the Ainu has been a point of contention among Japanese scholars. However, it is highly likely that Yamazaki assumed that the *ezo* was the ancestor of the Ainu from the Meiji period because the term, *ezo*, was used to indicate the Ainu since the thirteenth century. See William Wayne Farris, *Heavenly Warriors: The Evolution of Japan's Military, 500–1300* (Cambridge, Mass.: Council on East Asian Studies, Harvard University, 1995), pp. 82–83.
22. Yamazaki, op.cit., p. 112.
23. For health conditions of the Ainu in the late nineteenth and early twentieth century, see, e.g., Fujino Yutaka, *Nihon fashizumu to yūsei shisō* (Kyoto: Kamogawa Shuppan, 1998), pp. 216–259.

24. Ibid., p. 111. For Japanese policy toward the Ainu during the Meiji (1868–1912) and Taishō (1912–1926) periods, see Enomori Susumu, *Ainu no rekishi* (Tokyo: Sanseidō, 1987) and Takagi Hiroshi, "Ainu minzoku e no dōka seiseku no seiritsu," in *Kokumin kokka wo tou*, ed. Rekishigaku Kenkyūkai (Tokyo: Aoki Shoten, 1994), pp. 166–183; Richard Siddle, "The Ainu and the Discourse of 'Race', " in *The Construction of Racial Identities in China and Japan*, ed. Frank Dikötter (Honolulu: University of Hawaii Press, 1997), pp. 136–157; David L. Howell, "The Meiji State and the Logic of Ainu 'Protection', " in *New Directions in the Study of Meiji Japan*, ed. Helen Hardacre, with Adam L. Kern (Leiden, New York, and Köln: Brill, 1997), pp. 612–634. For Tokugawa state policy toward Ainu over the issue of vaccinations, see Brett L. Walker, "The Early Modern Japanese State and Ainu Vaccinations: Redefining the Body Politic 1799–1868," *Past & Present*, May 1999, *163*: 121–160.
25. Yamazaki, "Niton, Ainu, Ryūkyū, oyobi Shina yon shuzoku fujin no gekkei ni tsuite," p. 112.
26. Ibid., pp. 110–113.
27. Ibid., pp. 112–113.
28. About this issue, see Murai Osamu, "1910 nen nêshon and narêshon: teikoku no katari/metsubō no katari," a paper presented at the annual meeting for the Association of Asian Studies, April 1996; and also *Nantō ideorogî no seiritsu* (Tokyo: Ōta Shuppan, 1995), pp. 164–166.
29. Yamazaki, "Nihon, Ainu, Ryūkyū, oyobi Shina yon shuzoku fujin no gekkei ni tsuite," pp. 118, 166.
30. Ibid., pp. 118–119, 157–158.
31. Ibid., pp. 119–120, 157.
32. Ibid., pp. 119–120.
33. Ibid., pp. 156–157.
34. Ibid., pp.119, 120.
35. Ibid., pp. 110–111.
36. Ibid., pp. 158–159, 162, 166, 194.
37. Ibid., pp. 157, 163, 167, 169.
38. Ibid., pp. 171–172.
39. Ibid., p. 172.
40. Ibid., pp. 172, 174.
41. Ibid., p. 174.
42. Ibid., pp. 172–173.
43. Ibid., pp. 174–175.
44. Ibid.
45. Ibid., p. 174.
46. Ibid., p. 175.
47. Ibid.
48. Ibid., p. 173.
49. Ibid., pp. 173, 175.
50. Ibid., p. 126.
51. Ibid., p. 126.
52. Ibid., p. 127.
53. See, e.g., Michael Weiner, "The Invention of Identity: Race and Nation in Pre-war Japan," in *The Construction of Racial Identities in China and Japan*, ed. Frank Dikötter (Honolulu: University of Hawai'i Press, 1997), pp. 96–117.

54. See Yuki Terazawa, Chapter Five, "The Role of the State, Midwives, and Expectant Mothers in Childbirth Reforms in Meiji and Taishō Japan," and Chapter Six, "Women's Health Reforms in Japan at the Turn of the Twentieth Century," in "Gender, Knowledge, and Power: Reproductive Medicine in Japan, 1690–1930," unpublished Ph.D. diss., UCLA, 2001. For the discussions of menstruation by Japanese obstetrician–gynecologists trained in European medicine, see, e.g., Kinoshita Seichū, "Fujin ni hitsuyō naru eiseijō no chūi," *Fujin eisei zasshi* April 1901 (137): 1–20.
55. Tomiyama Ichirō, "Colonialism and the Sciences of the Tropical Zone: The Academic Analysis of Difference in 'the Island Peoples,' " *Positions* 1995, *3* (2): 367–391; also, Tomiyama, "Sokutei to iu gihō: jinshu kara kokumin e," *Edo no shisō*, July 1996 (4): 119–129.
56. Shimoide Sōkichi, "Miru to Supensā: Meiji bunka ni oyoboshita eikyō ni tsuite," in Shimoide, op.cit., pp. 34–49.
57. On this issue, see in particular Shimizu Akitoshi, "Colonialism and the Development of Modern Anthropology in Japan," in *Anthropology and Colonialism in Asia and Oceania*, ed. Bermen and Shimizu, op.cit., pp. 115–171.
58. On the question of how the issues of "*jinshu*" and "*minzoku*" were discussed during the early twentieth century, the 1930s and the World War II, see, e.g., Kevin M. Doak, "Culture, Ethnicity, and the State in Early Twentieth-Century Japan," in *Competing Modernities: Issues in Culture and Democracy, 1900–1930*, ed. Sharon Minichiello (Honolulu: University of Hawai'i Press, 1998); Tessa Morris-Suzuki, "Debating Racial Science in Wartime Japan," in *Osiris* 1998, *13*: 354–375; Sakano Tōru, "Kiyono Kenji no Nihon jinshu ron: Daitōwa kyōwaken to jinruigaku," *Kagakushi, kagaku tetsugaku kenkyū*, *11*: 85–99; "Jinruigakusha tachi no 'minami': senzen nihon ni okeru mikuroneshia jin kenkyū wo megutte, Part I," *Kagakushi kenkyū*, January 1997 (200): 239–250; and "Jinruigakusha tachi no 'minami': senzen nihon ni okeru mikuroneshia jin kenkyū wo megutte, Part II," *Kagakushi kenkyū*, April 1997 (201): 9–18; Sumiko Otsubo and James R. Bartholomew, "Eugenics in Japan: Some Ironies of Modernity, 1883–1945," *Science in Context* 1998, *II* (3–4): 133–146; Sumiko Otsubo Sitcawich, "Eugenics in Imperial Japan: Some Ironies of Modernity, 1883–1945," unpublished Ph.D. diss., Ohio State University, 1998; Fujino, *Nihon fashizumu to yūsei shisō*, op.cit. (1998): Yuehtsen Juliette Chung, *Struggle For National Survival: Eugenics in Sino-Japanese Contexts, 1896–1945* (London and New York: Routledge, 2002).

5

Doctors, Disease, and Development: Engineering Colonial Public Health in Southern Manchuria, 1905–1926

Robert John Perrins

The Home Islands of Japan witnessed incredible changes during the Meiji era as Japanese society, politics, foreign relations, and industry were transformed and modernized. Physical manifestations of modernization, such as railways, factories, and Westernized urban landscapes, were not the only evidence of the changes taking place in Japan in the latter half of the nineteenth century. A new nationalism also arose during this period—an ideology that, in part, came to be related to extending Japan's presence abroad through the acquisition of colonies. In 1895, following the first Sino-Japanese War (1894–1895), Taiwan became Japan's first colony. This was quickly followed by the acquisition of the Guandong (Kwantung) leasehold in southern Manchuria in 1905 and the annexation of Korea in 1910. Within only a few decades of embarking on its modernization drive, Japan had emerged, by the turn of the twentieth century, as a growing imperial power in East Asia.[1]

As was the case in Japan proper, life in the new colonies was shaped and (re)defined by the Meiji project of modernization during the late 1800s.[2] While the other chapters in this volume explore the roles played by science, technology and medicine in the complex path by which Japan negotiated the development of a modern state and society, this essay shifts our attention outward and explores the modernization process in a colonial setting—the Guandong leasehold on the Liaodong Peninsula in southern Manchuria, and specifically the port city of Dairen.

A warm spring sun was shining on the morning of May 21, 1927 as scores of conference participants began to arrive at the new South Manchuria Railway (SMR) hospital in Dairen.[3] From the front steps of the hospital, perched on the northern slopes of Nanzan hill, the invited guests could observe a panorama of evidence of the colonial port city's success. Dairen's central circle (*Ōhiroba*)—from which all major streets radiated, and around which the neoclassical buildings housing the city's administration and banks stood—could clearly be seen two blocks below. In the distance, one could also view the bustling harbor that was the economic soul of Dairen. After filing into the hospital, attendees were ushered into the auditorium where Dr. Todani Ginzaburō, the Superintendent-General of the hospital, was waiting to present his opening address. Dr. Todani began his remarks by welcoming his audience to the new hospital, a facility that was hailed by not only the speaker, but also the region's colonial governors, as the most modern medical facility in Manchuria, and, in a polite acknowledgement to the Chinese doctors in attendance, second in continental East Asia only to the Beijing (Peking) Union Medical College.[4]

Dr. Todani was followed at the podium by Banichi Yasuhirō, the President of the SMR. In his address summarizing the history of the new facility, President Banichi concluded that: "Medical knowledge and art, however advanced they may be, must depend on perfect equipment to cure and to heal. This was how the new hospital was planned and erected."[5] Over the course of the next three days conference participants listened to the more than 50 presentations that demonstrated how such modern, "perfect" equipment would be used to advance scientific knowledge. Medical researchers presented their findings on a wide-range of topics from a review of a five-year study of perspiration by Dr. Kuno Yasushi of the Manchurian Medical College in Mukden (Shenyang), to research into the nutritional value of milk for infants by Dr. Suzuki Tasashi of Kyoto Imperial University, to a debate over the creation of a smallpox vaccine between teams of doctors from the isolation hospital and the new SMR hospital in Dairen.[6] In addition to the research presentations, conference participants were taken on tours of the new hospital complex. Members of the hospital's administration and staff eagerly directed their visitors through the gleaming patient wards, well-stocked pharmacy, surgeries and laboratories equipped with the latest instruments from Japan and Germany, and even the maintenance and ice-making facilities in the subbasement.[7]

During the afternoon sessions of the conference, participants were divided into two groups that were led on guided tours of the city. In an effort to demonstrate that the hospital was not the only evidence of the modernity and progress that Japanese rule had brought to south Manchuria, attendees were shown other manifestations of colonial development including the new Nisshin Company's oil mills, the port's dock facilities, the SMR's research laboratories at its workshop at Shahekou, and a new ceramics and glass factory in the city's industrial quarter.[8] After participating in both the conference and the guided tours, it would have been difficult for visitors not to have been impressed by the state of medical care offered by the hospital, or, in fact, by the larger efforts of the city's Japanese governors to develop a colony that reflected the modernity and potential of post-Meiji Japan.

Although the larger region of Manchuria (figure 5.1) would later emerge as the "promised land" of Japanese imperialism after the creation of the puppet-state of Manzhouguo in the early 1930s, earlier Japanese colonial

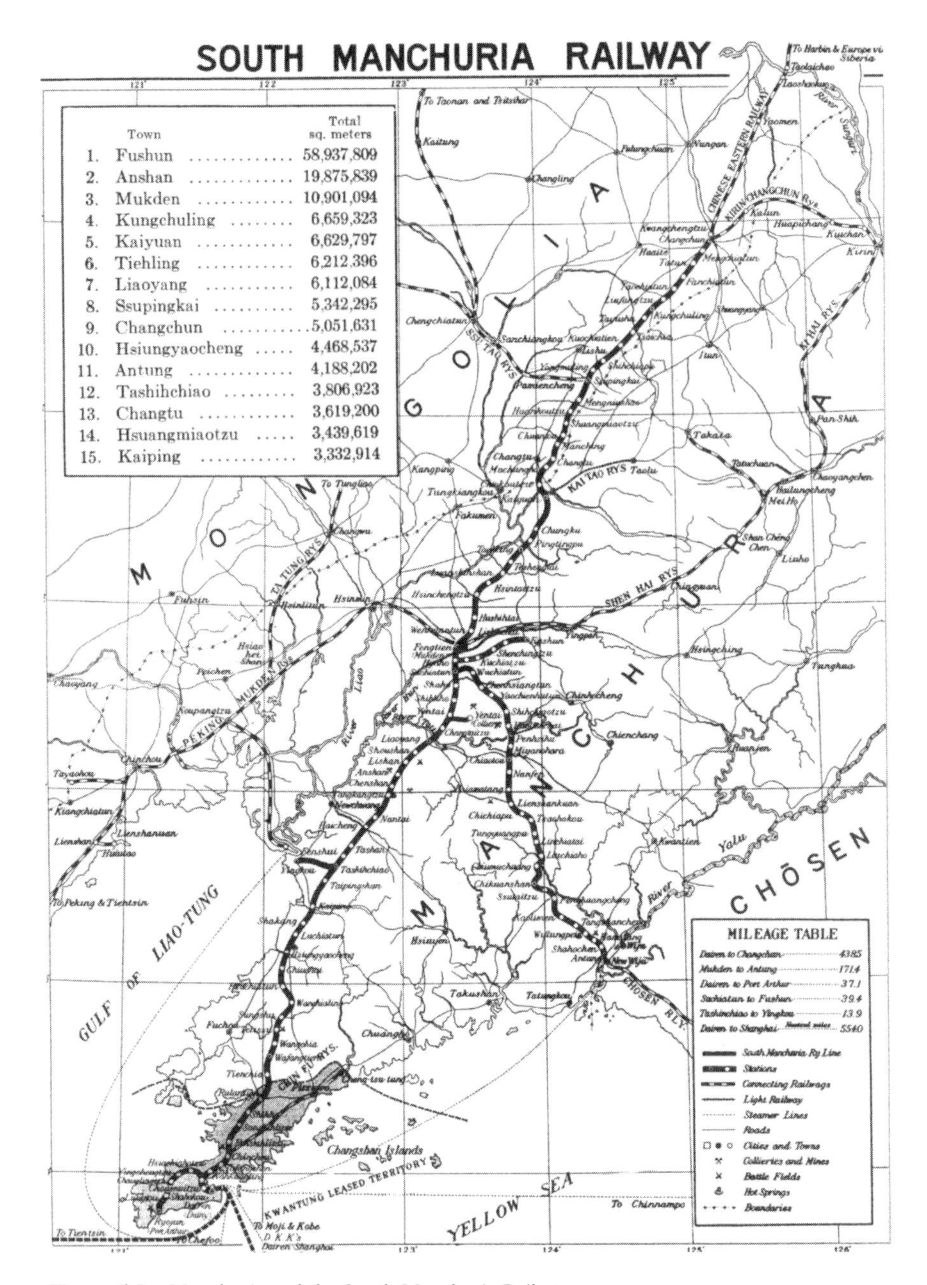

Figure 5.1 Manchuria and the South Manchuria Railway.

Source: Henry W. Kinney, *Manchuria Today* (Osaka: Hamada Printing, 1930), p. 101. The Guandong (Kwantung) leasehold is the shaded region at the tip of the Liaodong Peninsula in the bottom left corner of the map.

designs in northeastern China were focused on the smaller Guandong (Kwantung) Leased Territory, and the star of this early colonial performance was the home of the SMR's showcase hospital, the port city of Dairen.[9] Captured in the early months of the Russo-Japanese War, and formally transferred to Japanese rule with the signing of the Treaty of Portsmouth in 1905, Dairen grew to become the key administrative and commercial centre in southern Manchuria. During 1910s and early 1920s the Japanese governors of Dairen greatly enlarged the harbor and its wharves, oversaw the construction of a massive industrial quarter, and supervised the expansion of a city that they believed symbolised the "enlightened progress" brought about by Japanese colonial rule. By the 1920s, the leasehold's colonial governors were proudly proclaiming Dairen to be the most modern city in all of Manchuria.[10]

While many of the tools by which the Japanese were developing their colony in southern Manchuria would have been obvious to the participants of the 1927 conference—the derricks and cranes that dotted Dairen's wharves, the SMR's rail-lines and workshop at Shahekou, the city's factories, and even the medical facilities housed within the new hospital[11]—other, less obvious, tools may have been overlooked.[12] In order to protect their investments and to safeguard the health of the colonial population, the Japanese authorities in southern Manchuria had devoted a great deal of effort to creating technologies that aimed to improve the social or public health of the Guandong leasehold; these efforts ranged from city planning to building water reservoirs and sewers to enforcing quarantine procedures on arriving Chinese migrants.[13] Since the mid-1980s, there has been an explosion in scholarship that has examined the relationship between medical issues, and particularly Western medicine and its treatment of disease, and colonialism in Asia.[14] English-language works on the history of disease and medicine in the Japanese empire, however, have only recently begun to appear, and, to date, these works have focussed on the experience in Taiwan.[15] This chapter serves as a point of comparison both to the recent works on the colonial experience in Taiwan and to the other chapters in this volume that deal with the Home Islands, and examines how the "colonial state" in southern Manchuria attempted to fight disease and in the process create a "healthy" enclave that symbolized the "civilization" that accompanied Japanese rule.

One of the early areas of concern for the Japanese governors in southern Manchuria was the public health of the region, responsibility for which was initially assigned to the local military authorities. Only days after the signing of the Treaty of Portsmouth in December 1905, Lieutenant-General Kamio Akira, the commander of the Liaodong garrison and acting military governor of the region, ordered the establishment of a sanitation committee in Qingniwa—a small garrison town on the site of what would eventually be the northern districts of Dairen. This 38 member committee was staffed entirely by officers of the occupation army in southern Manchuria, and within a month it had established a number of regulations aimed at preventing the spread of contagious diseases in the counties surrounding the ports of Dairen and Ryojun (Port Arthur, Lushun).[16] Emergency edicts were published regarding the disposal of garbage, the supervision of the flourishing prostitution trade, and the establishment of mandatory quarantining of all new arrivals to the region.[17] These regulations aimed to prevent outbreaks of diseases that often accompanied the garrisoning of large numbers of soldiers in the region's ports while they waited to be shipped home.

In the years following the end of the war, the military's sanitation committee gradually evolved into a Public Sanitation Bureau that was charged with a broader mandate of ensuring "public" health throughout the leasehold. At first this bureau was a division within the Guandong Military Government (*Kantō Sōtokufu*) (1905–1906), and, after the restructuring of the territory's administration, it was reconstituted as a department within the Guandong Government-General (*Kantō Totokofu*) (1906–1919).[18] As the Kwantung Army (*Kantō-gun*) assumed greater control over the region's defense, the composition of, and control over, the sanitation bureau gradually shifted away from the military and into the hands of the civilian bureaucrats of the Guandong Government, and in larger towns and the port city of Dairen, the local constabulary.[19] Between 1907 and 1915, the Public Sanitation Bureau, together with the leasehold's Marine Quarantine Authority oversaw much of the early development of the healthcare system in southern Manchuria. Small local hospitals and dispensaries were opened in Dairen and Ryojun, including one administered by the South Manchuria Railway Company—often referred to as *Mantetsu*, an abbreviation of its full Japanese name (*Minami Manshū Tetsudō Kabushiki Kaisha*). The efficacy of these early facilities, however, proved inadequate in coping with the problems associated with urban growth in southern Manchuria, most notably in the port city of Dairen. As the populations of both the leasehold and city grew, so too did the rates of infectious diseases, particularly smallpox, typhoid and scarlet fevers, cholera, dysentery, and tuberculosis.

When the new Japanese governors of Dairen had arrived in the fall of 1905, they had been impressed by the future possibilities for the city and port, if not by the condition in which they found the Russian Finance Minister, Sergei Witte's "dream town." In the aftermath of the recent war, housing and most other buildings were in varying states of disrepair, and the streets were a disaster, impossibly dusty during the summer and impassable mud pits in the spring and fall.[20] While the SMR supervised the paving of the city's roads between 1908 and 1910, the Company's engineers and architects assisted the Guandong administration in planning the reconstruction of Dairen.[21] Fortunately, the Russian plans had been captured in May 1904, when the port fell to General Nakamura's Third Division, allowing the *Mantetsu* engineers to limit their work to modifying the original blueprints.[22]

The original Russian plans for their city of Dal'nii (Dairen) had been drafted in 1899 by Vladmir Sakharov, the chief construction and planning engineer for the Chinese Eastern Railway (CER).[23] Sakharov's ambitious plans had entailed a city of more than 80 hectares and a wharf capable of berthing more than 100, 1,000-ton vessels at a time.[24] Between 1899 and 1903, the Russians, at the insistence of Finance Minister Witte and to the annoyance of local military commanders, spent more than 30 million rubles developing the harbor and city.[25] The CER was charged with overseeing the construction of Dal'nii. Sakharov's blueprints, loosely based on the urban planning ideas of the "Garden City" and "City Beautiful" movements that were influencing the construction of cities throughout Europe and America, called for the new town to be built around five connected districts: one commercial, one administrative, two Russian residential, and lastly a "Chinese town" located a "safe" distance away from the other quarters. When the city was completed, the various sectors were to be linked by a spider's web of treed avenues, and all would eventually be supplied with electricity and a modern water system. Between 1899 and 1903, Russian engineers supervised thousands of Chinese laborers in building Dal'nii's main roads, rail lines, waterworks, and concrete wharf. Although the town failed to shine as brightly as Witte had promised, a start had been made, and during this initial period of construction, the port grew from a small fishing hamlet to a town of more than 35,000 persons.[26] By the start of the Russo-Japanese War, Dal'nii's Russian governors had begun to lay out what could be done, and although they had to a large degree abandoned the town by the time of its capture on May 30, 1904, they had completed much of the preliminary work of building a modern harbor.

In some ways, the work of the colonial planners in southern Manchuria foreshadowed later developments in urban design in the Home Islands. After 1905, the urban plans for Dairen, as modified by the *Mantetsu* engineers, continued to rely heavily on the original Russian theme of the "Garden City" model, a movement that would eventually have admirers among urban planners working in Japan during the late 1910s and 1920s.[27] As early as 1906, however, when the region was still under the direct control of the military, Japanese planners in southern Manchuria were already proclaiming that they were determined to develop Dairen into a modern colonial showcase. Kuratsuka Yoshio, an engineer, was appointed by the Guandong Military Government in early January 1906 to oversee the (re)construction of Dairen. In his preamble to the new ordinances that would regulate all construction in the city for the next 13 years, Kuratsuka wrote:

> Dairen is the base of [Japan's] management of Manchuria, and therefore, the size of the planned city should be both immense and modern so that we will not be ashamed of it in the future when the world observes what we have developed. Buildings should be insulated, fireproofed, and beautiful to look at . . . they should last forever.[28]

Not only was the physical appearance of buildings important to the early Japanese urban planners in Dairen, but so too was the physical "health" of the city and its residents—primarily, but not exclusively, those in the growing local Japanese community. Public health concerns were a central component in the overall colonial vision for the development of Dairen. The continued division of the city into linked, but separate, districts as originally proposed by Sahkarov's plans was directly related to the desire on the part of the new colonial authorities to create segregated Japanese and Chinese residential neighborhoods, as it was claimed that disease was more prevalent amongst the Chinese migrant laborer, or "coolie," communities, and therefore better controlled if limited to certain areas of the city.[29] It was also intended that Dairen would be a "green city," with numerous public parks and shaded avenues, as it was argued that the open air and parklands were part of the overall scheme in any modern, sanitary urban space. In fact, one of Dairen's most famous characteristics during the colonial era was the Acacia tree, thousands of which were planted during the first decade of Japanese rule. Clean air, clean paved streets, and segregated neighborhoods were all early components in the Japanese vision for their modern and "healthy" colonial showcase in southern Manchuria.

The new city plans also called for construction efforts that targeted the development of the city's harbor, a commercial and administrative showcase downtown, and an industrial quarter in the west end.[30] By the early 1910s, much of the first stage of Dairen's (re)construction was complete. A couple of electric tramlines had opened for public use in September 1910, the new Guandong Civil Administration headquarters was almost finished, and the glorious Yamato Hotel overlooking the *Ōhiroba* was accepting reservations.[31] In less than a decade, Dairen was on its way to being transformed from what in the minds of many of its original Western residents and consular officials was a shabby port town into a city that the most colonial of them would have been proud to call their own.

The greatest potential threat to Dairen during this initial period of Japanese rule was the outbreak in northern Manchuria in the winter of 1910 of the most feared disease of all, the plague.[32] It was somewhat ironic that the plague arrived in Dairen in December 1910 likely via brown rats that were transported in sacks of soya beans along the very railway that was responsible for the port's development.[33] Doctors and staff at the city's hospitals worked furiously to confirm that the disease that was showing up amongst small numbers of the city's population of migrant Chinese laborers was indeed the plague. Once the new threat was identified by pathologists employed by the SMR as the pneumonic plague, the city's doctors petitioned the local administration to draft emergency measures. Faced with this medical crisis, the Guandong Government requested assistance from the national government in Tokyo, and created the Guandong Temporary Municipal Sanitation Bureau (*Kantō totokufu rinji bōrekibu*) in Dairen, where the majority of the region's medical facilities and staff were located.[34] This agency worked with the local police and fire authorities to establish mechanisms by which it hoped to prevent the disease from spreading. Empowered by the Guandong Government with extraordinary legal powers, the Bureau established a tent city on the outskirts of Dairen to house the port's population of Chinese sojourning laborers, who were thought to be "natural" carriers of the disease.[35] The city's population of several thousand laborers from Shandong who worked on Dairen's docks and in the rail yards were forced to spend much of the winter of 1910–1911 living in drafty canvas tents, "safely" isolated from the port's Japanese population.

Within the city, the Bureau worked with the municipal police to establish check-posts at all major street intersections. As they moved throughout the town, residents were inspected at these posts for signs of the plague by sanitation crews wearing full-body isolation gear made of heavy white canvas. Houses where infected victims lived were sealed off with galvanized iron sheeting, and then sprayed with gallons of concentrated formaldehyde. The fences of iron sheeting were used to prevent the city's rat population from simply moving from house-to-house, and even block-to-block, while poisonous gasses were pumped into the buildings suspected of housing the disease.[36] By the end of March 1911, the worst of the epidemic had passed. The tent city outside of Xiaogangzi was dismantled, and its Chinese population sent back to their homes or dormitories; and the sanitation carts that were filled with drums of formaldehyde and rat poison were put into storage, hopefully never to be used again. At the final count, when the last cases of the disease were recorded in the summer of 1911, a total of 5,864 people had died of the plague in southern Manchuria (including 2,495 in Changchun and 2,005 in Mukden [Shenyang], but only 66 in Dairen).[37] The death count for the region as a whole was staggering, with more than 60,000 having died throughout Manchuria.[38] It should be noted, however, that all of the 1910–1911 plague victims in colonial southern Manchuria were Chinese.

It was the quarantining of the entire city of Dairen, through the closure of the port to railcars from the north, and not the isolating, or perhaps more accurately imprisoning, of the city's Chinese laborers that ultimately prevented the plague from taking hold on the Liaodong Peninsula. Because the Guandong Government was willing to risk the wrath of the leasehold's commodity traders and merchants, who were admittedly angered at the temporary closing of the railway lines to the north, Dairen survived the great Manchurian plague epidemic of 1910–1911 relatively unscathed. In the years that followed, even more trade flowed into the port, and its economy prospered. To the delight of both the region's merchant community and the colonial administration, the economy of southern Manchuria blossomed in the late 1910s due to the growing world demand for soya beans and their products. Between 1913 and 1917, tonnage of raw beans exported to these new markets through the port of Dairen grew 50 fold, from under 5,000 to over 250,000 metric tonnes. The growth in the oil trade during this same period was equally dramatic, with the export to the United States alone growing from just over 3,000 to almost 100,000 tonnes.[39] To handle the increase in the traffic of soya beans and bean products the Japanese governors of Dairen had to react swiftly and find a way of constructing additional warehouses and wharf facilities, for by the late 1910s more than 55,000 railcars filled with beans were arriving annually in the port for processing and shipment abroad.[40]

During the economic boom of the late 1910s, the population of Dairen grew dramatically, growing from just over 120,000 persons in 1914 to almost 200,000 persons by the end of the decade.[41] In an effort to cope with this tremendous demographic growth, the Guandong administration developed a set of new urban plans for Dairen. The original Russian plans had detailed a city of approximately 80 hectares, and following the transfer of the Guandong leasehold in 1905, the new Japanese rulers had spent vast sums developing their colony. Within a couple of years, the region's railway system had been re-gauged, and new engines, cars and other equipment imported from America. To service this improved transportation system, and to connect it to the outside world, the region's Japanese governors focused their efforts on the construction of the leasehold's primary commercial port at Dairen. In 1910, the urban plans for Dairen had been expanded to encompass a city of 700 hectares—almost a ten-fold increase in size from the city sketched in the original Russian plans.[42] By the late 1910s, however, with the growth of local industry and the port's population, even the revised blueprints were obsolete as Dairen was quickly growing beyond its planned boundaries. One of the first tasks of the municipal government that was elected for the first time in 1915, therefore, was the development of a new urban plan that would see the city into the middle of the twentieth century.

The city plans that were developed between 1915 and the spring of 1919 were extensive (figure 5.2), detailing a city of over 2,000 hectares and a harbor with six concrete wharves—a number that was three more than the SMR had

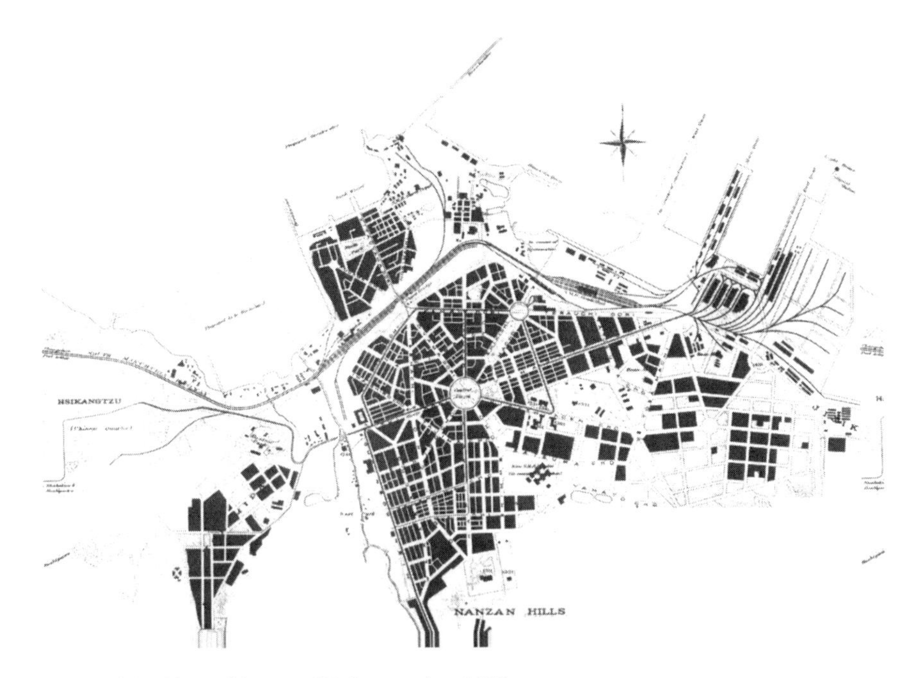

Figure 5.2 Planned layout of Dairen, ca. late 1910s.

Source: *Dairen (Dalny)* (Dairen: Japanese Tourist Bureau, Dairen Branch, 1917), end map. The area that comprised the original Russian town of Dal'nii is located in the upper middle portion of the growing city, to the west and southwest of the harbor's main wharves.

so far completed.[43] Under the joint supervision of the SMR, the local wharf authorities, and the new municipal government, Dairen was emerging as one of the most impressive trading centers in Asia. During the planning for the port's future, public health was an important element in the colonial vision of Dairen's development. The continued division of the city into linked districts as proposed by the "Garden City" model that was now in vogue in Taishō Japan was related to a desire on the part of the authorities to create segregated Japanese and Chinese residential neighborhoods for reasons that were based on a "scientific" understanding of where diseases flourished (i.e., the less "hygienic" Chinese community was thought to be a "natural" incubator of illnesses).[44] Throughout the first two decades of Japanese colonial rule in Dairen, city planners and administrators went so far as to draft and enforce zoning regulations that actively sought to exclude Chinese from living in what were designated to be Japanese neighborhoods. Along with "clean" residential quarters and green parks, the revised city plans also called for the expansion of the harbor facilities, the completion of the commercial and administrative showcase downtown, and further additions to the industrial quarter in the west end. These parts, or organs, of the body of Dairen were to be linked with a "modern" nervous system—an expanded system of electric tramlines that ran through the city-centre and out into the developing western and southern peripheries. In 1929, the Guandong Government published an English-language book that detailed its history and responsibilities. In this book the contribution of the colonial administration to the growth of Dairen was highlighted: "Everything has been done to develop the city, making the most of the available land and to keep up a high hygienic standard with an efficient police service."[45] By the 1920s, Dairen was on its way to becoming a beacon of Japanese "enlightenment" and "civilization" in the growing empire—a port that was not only an economic success, but also a showcase of colonial urban planning and hygiene.

As outlined above, trade was not the only thing that flowed across Dairen's wharves during the first two decades of Japanese rule, as shortly after the conclusion of the Russo-Japanese War, tens of thousands of Japanese began arriving in the Guandong territory to seek their fortunes. In 1906 alone, over 30,000 Japanese civilians had disembarked at the newly opened port of Dairen. The British Ambassador in Tokyo at the time, Sir Claude MacDonald, reported that many of these Japanese were "adventurers possessing little or no money," or women, "notoriously of the lowest type," who were there to service the withdrawing military forces.[46] Scores of commercial traders from Japan and China also flocked to Dairen during the early years of Japanese rule, their boats filled with bolts of cotton textiles, drums of kerosene, and boxes of foodstuffs and other consumables.[47] The largest category of new arrivals, however, were the tens of thousands of sojourners from Shandong Province who came in search of employment on the city's docks and construction lots. Fueled by so many new arrivals, Dairen's population grew dramatically in the first two decades of Japanese rule, increasing from 38,896 in 1906, to 278,545 by 1926.[48] In an effort to prevent any of the new arrivals from transporting infectious diseases to the city's increasingly crowded neighborhoods, the Guandong authorities, together with the SMR, established rigorous quarantine procedures at Dairen's wharves.

Established in November 1908, the Marine Affairs Bureau of the Guandong Government was charged with supervising all quarantining procedures at Dairen.[49] This agency developed a rigorous system of medical inspection for vessels arriving in the harbor, and built a large quarantine facility just south of the main piers to house hundreds of new arrivals who had been deemed to be possible carriers of disease by the bureau's doctors. Inspectors from the bureau examined new arrivals for signs that they were carrying one (or more) of the 12 reportable infectious diseases: plague, cholera, smallpox, typhus, scarlet fever, diphtheria, dysentery, typhoid, sub-typhoid, encephalitic fever, encephalitis B, and general fever. Ships sailing to Dairen had to radio their passenger and crew manifests, and the passengers' medical conditions one day prior to arrival at the harbor. If a vessel did not have radio equipment on board it had to fly a yellow flag upon its arrival within the harbor and await a team of medical examiners. The doctors and assistants who worked for the Marine Affairs Bureau would then sail out to the waiting ship and give each passenger a quick medical examination.[50] There were two categories of incoming vessels: (1) those from Japan, Korea, Taiwan, Europe and America; and (2) those coming from ports in China and other parts of Asia that were not Japanese colonies. Ships from the first category generally had a quick passage through the quarantine procedures, and only those passengers who exhibited signs of infection were forced to undergo an additional quarantine period in special isolation facilities located adjacent to the harbor. For ships from China and the rest of Asia the process was much more rigorous, and all passengers, whether or not they exhibited signs of infection, were forced to undergo mandatory sanitary showers with diluted formaldehyde before being issued with their disembarkation passes.[51] Although blatantly discriminatory in nature, the port quarantining policies that were enforced at Dairen during the colonial period, did, to an extent, prevent the arrival by ship of numerous, potentially serious, diseases.

After the rapid growth of both the soya bean trade and the port city of Dairen in the late 1910s, the local Japanese authorities developed additional, and more invasive, public health policies and enforcement agencies to protect the health of their goose that was laying the proverbial golden eggs. The construction of mechanisms for tracking and combating diseases in Dairen was a central component in the continuing development of the colonial infrastructure in the most important urban space in the Guandong leasehold. Dairen was not only a port, and therefore susceptible to diseases that could arrive either via ships or from the region's railway network, but also a large and growing urban landscape that was now home to hundreds of thousands of residents. A result of the efforts to scientifically study and treat disease in Dairen was the creation of what, in another colonial context, has been termed a "pathogenic city."[52] In her study of British colonial medicine in

Singapore and Malaysia in the early twentieth century, Lenore Manderson notes that towns and cities were:

> [T]he most domesticated space, contrasting with the untamed, unpenetrated hinterland and the diseases that it harboured. The town reflected the colonists' best efforts to create colonial space out of, and in face of resistance from, place, and issues of the use and misuse of urban space were examples of their imperfect control over both place and people. The prevalence of diseases . . . highlighted and constituted part of the difficulty of establishing territorial control. But at the same time, use of space and human interactions with and use of both the natural and built landscape reflected the cultural tensions of colonialism. Obsession with hygiene and sanitation in the cities was part of this.[53]

To meet their "obsession" with hygiene, and to keep the pathogens at bay, the Japanese governors devised a myriad of agencies and levels of bureaucracy that collaborated in establishing and enforcing public health policies in southern Manchuria. At the top of the colonial hierarchy stood the Guandong Governor-General and his administration, and operating under this lofty level of colonial government were the municipal government, the local constabulary, the Maritime Affairs Bureau, and several local sanitation bureaus.[54] During the late 1910s and early 1920s, the Guandong administration passed a number of public health bylaws that were strenuously enforced by the local constabulary and health officers in Dairen. Among these new laws were regulations governing mandatory vaccinations for the local population, public health codes for restaurants and other food vendors, health curriculum for government-run schools, water monitoring and inspection codes, and rules governing the collection and disposal of "night soil." To support this new campaign against disease, the Guandong Government and the Dairen District Health Agency, a division of the local Dairen municipality, devoted between 10 and 15 percent of their annual budgets to medical and sanitary expenses.[55]

Although some of its responsibility for supervising the leasehold's public health had been transferred to the Dairen District Health Agency after the city's administration was revamped in 1915, Japan's greatest colonial enterprise, the South Manchuria Railway Company, continued to play a central role in "colonizing" the health of the local population throughout the second decade of Japanese rule.[56] Together with the Civil Administration Department of the Guandong Government-General and the Dairen municipal government, the SMR oversaw much of the development of Dairen during the first two decades of Japanese rule.[57] In addition to operating several hospitals along the railway zone and in the towns and cities in the region, *Mantetsu* was also charged with supervising local water supplies (including monitoring their qualities), as well as running a medical laboratory at its central experimental and research station in Dairen. While the port city was hailed in *Mantetsu* advertisements, and by the local colonial authorities and chamber of commerce to be a "city of lights" and a "beacon of civilization," behind the bravado lurked a fear that the city, and particularly its growing Chinese population, was incubating diseases that could strike at any moment tarnishing not only Dairen's image, but also the profits that were being made in the leasehold's transportation nexus.[58] As was the case in other colonial ports such as Singapore, public space in Dairen reinforced the idea of "a pathogenic city which characterized much of the medical thinking at the time . . . illness was produced in certain enclaves";[59] enclaves that on the Liaodong Peninsula included Dairen's harbor, regional asylum and isolation hospital, the port's overcrowded Chinese residential quarter, dormitories, slums, and "unsanitary" businesses such as brothels, markets and restaurants.[60] Throughout the first two decades of Japanese colonial rule in southern Manchuria, the Guandong administration worked with local authorities and the *Mantetsu* hydra to construct a sanitary colony. In the attempt to prevent disease from arriving in Dairen, vigorous quarantine procedures were enforced in the harbor, while within the city the local police and sanitation workers patrolled and inspected neighborhoods and businesses, keeping records on all transgressions of public health policy. By the late 1910s, the fight against disease had become a key component in the development of the growing Manchurian colony.

Of course, no matter how diligently the enforcers of public health worked, diseases did manage to navigate their way to Dairen. The warm summer and autumn months in particular witnessed the annual return of deadly pathogens that were easily spread by human locomotion and which thrived in poor sanitary conditions. As the temperature in the port increased, the city's residents had to contend with regular outbreaks of dysentery and other microbial illnesses of the digestive system, which claimed upward of 3,000 residents annually.[61] To combat the yearly onset of deadly stomach ailments, local police and health inspectors in Dairen stepped-up their inspections of the city's restaurants, water supplies and waste collection facilities. During serious outbreaks of dysentery, members of the city's constabulary were often ordered to close Dairen's markets to fish and vegetables that were caught or grown locally. Despite efforts to enforce sanitation ordinances, mortality rates for both dysentery and diseases/infections of the digestive track did not appreciably decrease during the first two decades of Japanese rule with the rate for dysentery ranging between 5 and 25 deaths per 100,000, and figures for the more general category of digestive system ailments fluctuating between 240 and 470 deaths per 100,000 persons.[62] The lack of efficacy of the efforts to eradicate the microbial threats that faced Dairen can partly be explained by the fact that, despite the enforcement of public hygiene bylaws, the port lacked an adequate supply of clean water that would allow its water and sewer systems to function to their full potential.

Seasonal illnesses were not the only ones that stalked the streets of Dairen during the 1910s and 1920s. In addition to the annual arrival of summer stomach ailments, the city's residents also faced other health problems. During the early twentieth century, as was also the case in Japan (and in many other major urban centers throughout the world), tuberculosis was the constant "plague" in the port city.[63] This disease was highly contagious not only in the Chinese workers' dormitories, but also in the supposedly sanitary

Japanese residential and commercial districts due to the city's generally crowded conditions. The mortality statistics for tuberculosis for the Guandong leasehold for the late 1910s reveal that it was a serious problem in the growing colony, accounting for almost 1,500 deaths per year.[64] To meet the challenge of the "white plague," local authorities in Dairen worked with the SMR both in developing a public health campaign that aimed to educate the local population on how the disease was communicated, and in constructing a large sanitarium where those who were suffering from the consumption could be treated.[65] While medical facilities were built to treat those who had contracted tuberculosis, it must be remembered that yet again the only ones who were to benefit were the leasehold's Japanese residents, as Chinese patients were not permitted to seek treatment in the sanitarium.[66]

At the end of the 1910s the economy and physical health of the Guandong leasehold and its inhabitants were further threatened by two epidemics. The first was the arrival of the "Spanish Lady" in 1918–1919. A mild strain of influenza had passed through the region in the spring of 1918, resulting in thousands of inhabitants complaining of fevers, aches and general malaise, but relatively few deaths. In mid-October 1918, however, reports began to appear documenting the return of the "flu" amongst military personnel stationed at the naval base in Ryojun (Port Arthur). By the end of October the population of Dairen found itself facing a major influenza epidemic that was much more serious than the authorities had at first believed.[67] Over the course of the next two months, municipal authorities enacted a number of provisions in an attempt to combat the strange illness that was paralyzing not only the leasehold, but also much of the world.[68] The emergency public health measures that were introduced in southern Manchuria in the fall and winter of 1918–1919 were similar to those that were used in the Home Islands, and involved the closing of public places such as schools, theaters, bath houses, swimming pools, and markets; the ordering of the population to wear gauze masks when in public; and new rules that forbade spitting or coughing in public.[69] The problem was that many of these regulations were passed too late to be of much use, as the virus had already circulated among the local population. When the final figures were compiled the following year, authorities claimed that up to one-third of the local population had been infected, and of these 3,354 had died as a direct result of the influenza virus.[70]

Those who survived the flu epidemic were faced with the arrival of another, even more frightening disease the following year. The disease that posed the greatest potential threat to Japan's "sanitary colony" in Manchuria was cholera, and the dangers associated with this illness were greatest in the densely populated port of Dairen. In late August 1919, as the city was only just beginning to recover from a third, and final, wave of the influenza pandemic, cholera arrived. The local authorities had known that the disease was making its way along the Chinese coast, and had taken the preventative measure two weeks earlier of hiring an additional 100 police constables and 18 doctors from Japan to assist in the quarantine inspections at the harbor.[71]

Within the city, police and sanitation officers enforced the closing of all produce and fish markets, and additional teams of municipal workers were dispatched to remove garbage and pools of stagnant water from the city's streets and neighborhoods. After a corpse was found floating in the city's reservoir during the second week of the epidemic, armed police were stationed at the facility to prevent the further contamination of the port's water supply.[72]

The staff at the Dairen Isolation Hospital was put to the test during the month-long crisis, as not only did they have to treat several hundred cholera patients, but they were also charged with vaccinating the port's Chinese laborers, who it was claimed were "dead to all sense of public and individual hygiene."[73] In an effort to "sanitize" the port's Chinese dockworkers, who presented a threat perceived to be as great as the cholera bacillus itself, teams of doctors and nurses spent the first two weeks of September inoculating more than 9,000 sojourning laborers with an anti-cholera vaccine. By the end of the first week of September it was clear that despite the efforts of the local constabulary, doctors and nurses, the battle to contain the epidemic was failing, and the Government was forced to request that the Foreign Ministry send an additional 152 police constables and 52 physicians from Japan to assist the beleaguered local public health workers.[74] With the assistance of these additional forces, the tide eventually turned and the last new case of cholera was reported on September 28. By the end of the outbreak, and despite all of the efforts to protect the "sanitary" colony, more than 1,600 persons, the majority of whom were Chinese laborers who lived in the crowded dormitories near the dockyards, died during the 1919 cholera epidemic.[75]

Although it appeared seven years after the 1919 epidemic, an editorial cartoon in the September 1, 1926 edition of *The Manchuria Daily News* illustrated the fear that cholera continued to hold in the minds of the port's residents. In the cartoon, a menacing cloaked figure threatens a cowering citizen who holds a microscope, a seemingly innocuous instrument of modern medical science, that he points like a revolver at the dark figure labeled "Bacilli." Although there was a general belief, at least among the region's Japanese and small Western communities who enjoyed access to modern medical care, in the ability of medical science to defeat the microscopic bacteria and viruses that threatened the colony in southern Manchuria, there was also the realization that cholera and other diseases that flourished in unsanitary conditions would only be eliminated if the city's water supply system was drastically improved. Regular outbreaks of diseases such as dysentery, typhus and smallpox, combined with serious epidemics such as the influenza pandemic of 1918–1919 and the cholera outbreak of 1919 served to heighten administrative concerns regarding urban sanitation, port quarantine, and water quality in Dairen. During the early 1920s, local administrators focused much of their energy on securing adequate supplies of clean water and modern sewage systems for Dairen. Although the city's planners had envisioned a "garden city," the reality was now an increasingly crowded town that was often dusty, malodorous and parched in the summer months. Beginning in the late 1910s,

Dairen faced annual water crises in July and August when low water levels in the town's reservoirs meant that the city's sewage system often failed to work at maximum efficiency.

Located on a rocky peninsula with no available groundwater, Dairen, with its annual rainfall of just over 600 mm, had always been poorly situated, in terms of water resources, to develop into a major city. In fact during the first couple of years of Japanese rule, municipal engineers attempted to bring water to Dairen from a reservoir at the town of Dayukou near the harbor at Port Arthur, 30 kilometres away, using a combination of water tenders and railcars. When these dramatic efforts failed, test wells were drilled between December 1912 and July 1913 at a number of locations throughout the city in an unsuccessful search for water.[76] Following the failure to locate sources of groundwater in the vicinity, the Guandong Government decided in 1914 to expand the newly completed Shahekou filtration plant and to bring in additional water from an enlarged reservoir at Wangjiatian, 20 kilometers northwest of the city. Construction of this great reservoir, capable of holding more than five million tonnes of water, and at a heady cost of almost two million yen, was completed between April 1914 and August 1917.[77] To commemorate their "successful" conquest of the city's water crisis, the SMR and the Guandong administration ordered 2,000 cherry trees to be planted along the banks of the reservoir that would now guarantee Dairen's future sanitary and commercial growth.[78]

Although the new reservoir theoretically should have solved Dairen's water crisis, the problem persisted for three reasons. First, the reservoir was rarely filled to its capacity due to poor precipitation in the region and heavier than expected industrial usage in the port. Second, the old reservoir was often contaminated by human waste from run-off from the surrounding farms that were fertilized with "night-soil."[79] Third, the city had grown beyond what the planners had forecast in the mid-1910s.[80] The growth of Dairen's population and industrial sector during the soya bean boom of the late 1910s, together with the cholera outbreak of 1919, combined to bring about louder calls for additional supplies of clean water. By the early 1920s the Guandong administration and Dairen's municipal government faced increased pressure from the port's chamber of commerce and local residents to secure clean water for both commercial and personal needs, and once again the colonial bureaucrats turned to the SMR for technical and financial assistance.[81]

A decade and a half earlier, *Mantetsu* engineers and construction crews had built Dairen's first modern waterworks.[82] Located in Shahekou (the SMR's city within a city in the western-most district of the port), Dairen's earliest filtration plant and waterworks were situated on the Malan River, seven kilometers west of the city center. Water from the 30 kilometer-long Malan, as well as the Wangjiatian Reservoir, was treated at this facility run by the SMR. Between 1905 and 1910, the Guandong authorities, together with *Mantetsu*, spent more than one million yen constructing the water treatment plant and pumping station, as well as the reservoir's earthen-works and

20-meter high dam.[83] A subsidiary reservoir was then built in the summer of 1910 in the Japanese residential neighborhood of Fushimidai, just west of the city center, and water from both locations was brought into the city proper through a network of 50 cm diameter pipes. In an effort to cope with increased demands for fresh water, the Malan filtration plant was expanded in 1914 at an additional cost of 1.3 million yen. The opening of this facility had been officiated by not only members of the city administration and representatives of the SMR who designed and built the treatment plant, but also by a local Shinto priest who blessed the equipment before the pumps started the flow through the city's main pipes.[84] This sanitized water supplied the "modern" private homes in Dairen's Japanese residential quarters, as well as the city's fire hydrants, bathhouses, public parks, industries, and harbor. The city's Sanitary Union also had a contract that allowed its members to open the city's hydrants and sell water, at the rate of two sen for each two gallon can that was filled, to residents who did not have indoor plumbing.[85]

Despite these efforts, by the early 1920s the port of Dairen continued to have no reliable source of clean drinking water, and therefore another attempt was made to find a "scientific and rational conclusion,"[86] to the city's water crisis. In his hagiographic article on the history of the Dairen waterworks, Dr. Y. Kuratsuka, the director of the Dairen Public Works Department wrote:

> [In the] city of Dairen, the surface water, meager as it may be, must be relied upon. And, unlike other cities in Japan and elsewhere, where natural sources such as rivers, lakes, etc., are available, what may be styled the reservoir system is the only practical way open.[87]

Kuratsuka went so far as to claim that the extension of Dairen's waterworks was a "problem of life and death to the growth of the city," as water was vital to both the port's economy and the health of its residents.[88] In their continuing attempt to create a stable and safe water supply for the port, SMR engineers went back to their drafting tables in 1921 to plan expansions of the existing Wangjiatian reservoir and the treatment facility near Shahekou, as well as the construction of a new reservoir at Longwangdang, 25 kilometers west of the city.[89] When completed in 1924, this new reservoir had cost the colonial administration more than four and a half million yen to build. Even before the first shovel-full of earth was moved at Longwangdang, however, a more cautious group of engineers and public works officials had already recognized that even this new facility would not be able to quench the thirst of the constantly growing port.

Throughout the decade of the 1920s, and into the 1930s, securing adequate sources of clean water for industrial and human needs continued to be one of the greatest problems facing Dairen's colonial governors. As thousands of new residents arrived each year to settle in the "garden city" additional strain was placed on an already stressed waterworks. Colonial planners estimated that, apart from the needs of local industries and the thousands of vessels that visited the harbor annually, tens of millions of litres of clean water were needed to service the city's growing population. In their calculations, the Japanese municipal planners and SMR engineers estimated that each of Dairen's Japanese (and small numbers of Western) residents required 4.5 cubic feet of water daily, while the "less hygienic" Chinese residents could cope with only one cubic foot per day.[90] As Dairen grew during the first two decades of Japanese rule, the search for a stable supply of clean water remained a key concern of the colonial administration. In the name of public health and urban sanitation, engineered technologies worked hand-in-glove with colonial medical concerns, and were, despite their failed efforts, hailed as critical components in the "scientific" development of the Japanese colony in southern Manchuria.[91]

The building and expansion of Dairen's various reservoirs, municipal sewer lines, and water treatment facilities were not the only major construction projects undertaken by the local Japanese authorities in an effort to safeguard the health of the colony in southern Manchuria. By the early 1920s, in the wake of the influenza and cholera epidemics of 1918–1919, and in the face of a large, and growing, urban population in the port city of Dairen, the Guandong Government and the SMR decided to build a new hospital. Construction of this facility was finally begun in March 1923, although preliminary plans had first been drafted as early as 1912, and the land purchased two years later in 1914. Once started, it took three full years to build and equip the new hospital. The resulting Romanesque edifice to colonial modernity (figure 5.3) was designed by architects at an American construction firm

Figure 5.3 The New South Manchuria Railway Hospital, ca. 1926.

Source: *Dalian shi tushuguan* (Dalian Municipal Library), historical photographs collection, and the author's personal collection.

based in Tokyo, George A. Fuller and Company of the Orient. The decision to award the three million yen contract for the building of the facility to a foreign firm was a departure from previous practice in Dairen that had favored Japanese construction companies.[92] At the time that the tender for this major contract was advertised, Tokyo was still shaking from the aftershocks of the Great Kantō Earthquake of 1923. Despite the fact that its staff had never before designed or built a hospital, Fuller and Company was ultimately awarded the contract for two reasons.[93] First, the firm had designed several large buildings in Tokyo, including a number of corporate headquarters and government offices in the capital's Marunouchi district, just northeast of Hibiya Park, that had survived the quake, resulting in an enhanced reputation for the company's ability to construct not only modern, but also solid facilities.[94] And second, just as *Mantetsu* had relied on American technical expertise and equipment when it rebuilt the region's railway network almost two decades earlier,[95] many members in the Japanese colonial administration continued to believe that Western scientific and engineering abilities were still slightly more advanced than their own.[96]

Between the spring of 1923 and the winter of 1925, the hospital's construction was completed by a team of almost 1,000 skilled and unskilled Chinese laborers, who worked under the supervision of Japanese foremen and American engineers.[97] Although the final cost was two million yen more than the initial estimate, the new hospital was completed on time, if not on budget. The first patients arrived for treatment and surgeries in April 1926, and entered what was referred to at a major conference the following May as a symbolic representation of Japan's enlightened development of Manchuria, and even a "mirage of grandeur."[98] The new hospital was, admittedly, an impressive facility comprising the main six-story building (including a basement and subbasement) (figure 5.4), a three-story isolation ward, two three-story

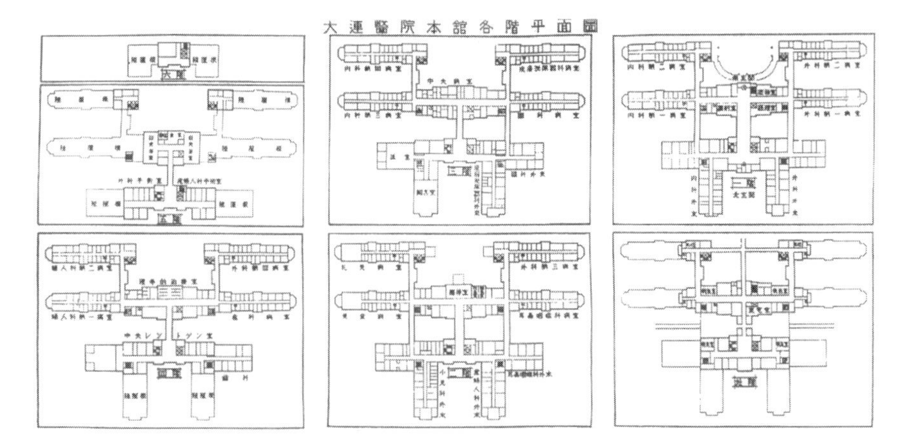

Figure 5.4 Floor plans of the New South Manchuria Railway Hospital in Dairen, ca. 1926.
Source: *Dairen Iin gaiyō* (an outline of the Dairen Hospital) (Dairen: n.p., 1927), frontpiece illustration no. 3.

nurses quarters and patient rooms, and an auxiliary equipment and maintenance building, all of which were connected through an underground network of tunnels.[99] The floor space of the facility was enormous, amounting to more than 45,000 square meters, making this hospital the largest medical facility in northern China. Stocked with the latest medical, pharmaceutical and laboratory equipment and supplies, the Dairen Hospital treated hundreds of outpatients daily, and had beds for almost 600 more in its inpatient wards.

Not all residents of the port city and surrounding leasehold, however, were to benefit from the facilities that could be found in the new hospital. The grand SMR hospital in Dairen had been constructed primarily for the treatment of the colony's sizable Japanese community, but even within this group, not all would have access to the "medical wonders" offered by the highly trained staff of doctors and nurses. Although its spokesmen often bemoaned the financial sacrifice that the SMR was forced to endure in providing health care facilities in the colony, the company, through the fees charged in facilities such as the Dairen hospital generated sizable revenues throughout the colonial period.[100] Daily room fees in the new hospital ranged from over ten yen for a "special first class room," down to two yen for a bed in a third-class ward.[101] Even the lower fee, it should be noted, was not trivial as the average weekly salary for even a skilled Japanese worker in the port was under five yen.[102]

While working-class Japanese faced a financial hurdle to seeking treatment in the new hospital, the situation was far worse for members of the city's Chinese community. Chinese employees of the SMR were permitted to seek basic treatment at the hospital, although they were placed in a segregated ward in the basement of the main building. Tens of thousands of other Chinese residents in the port, however, were never allowed to step into the red brick edifice to progress that overlooked their city. Several smaller clinics and hospitals were built to treat Dairen's Chinese majority, but these facilities could not offer the level of care provided by the staff and equipment at the main hospital.[103] In their efforts to construct a healthy colony, the Japanese governors were also constructing barriers not only between diseases and people, but also between ruler and ruled.

The port city of Dairen played a central role in Japan's early colonial efforts in Manchuria. For the experiment in the Guandong leasehold to be successful, Dairen had to be not only economically prosperous, but also, and perhaps as importantly, viewed by the outside world as a glowing symbol of the modernity that accompanied Japanese rule. An important element in the development of Dairen during the early twentieth century was the construction of the city's public health infrastructure. In this process, "purely" medical or health issues often interacted with broader economic and administrative interests. While not always at the forefront, as it was during the 1910–1911 plague outbreak, or the influenza and cholera epidemics of 1918–1919, the health of the colony in southern Manchuria was of vital interest to the Japanese colonial governors. This chapter has begun the examination of how medical concerns were often part of larger developments ranging from the physical layout of the city to efforts to engineer access to clean water. In the spring of 1926, just as the new SMR hospital was preparing to open its doors, the American journalist, R.O. Matheson passed through Dairen on a tour of Manchuria. He was impressed by what he saw in the city, and described his impressions of this port in an article that was published by his sponsors at the *Chicago Tribune*:

> Wide, paved streets, splendid hotels, handsome and substantial public buildings, a score of splendidly housed banks, produce and stock exchanges, theaters, cinemas and clubs make Dairen the best-built city of Manchuria. . . . A great hospital, the finest in all the Far East, with a thousand beds, ten or twelve operating rooms, four X-ray rooms and every appliance of modern medicine and surgery, has just been completed, at a cost of more than Yen 6,000,000. If Dairen is destined to become the city of refuge for foreign business in China, it will be a good one, clean, healthy and up-to-date.[104]

The tone of Matheson's report would surely have been music to the ears of the first generation of Japanese urban planners, public health officials, and colonial administrators in southern Manchuria.

Despite the grandeur of the Ōhiroba and the amenities available to the port's Japanese residents that impressed Western observers such as Matheson, however, the majority of Dairen's Chinese population continued to live in the shanty town that had grown up south of the port's wharves, or in Xiaogangzi—the growing Western neighborhood between the SMR's workshop in Shahekou and the city's downtown core. By the 1920s it was evident, if one explored all of Dairen, that two cities were being built: a modern, Western-styled municipality, dotted with trees and impressive new buildings, which was to be enjoyed by the port's Japanese community, and another more cramped and desolate in appearance that was to be coped with by the town's Chinese majority. As city streets, water-mains, tram-lines and architectural styles in Dairen physically divided residents into colonizers and colonized, so too did other seemingly less overt manifestations of colonial rule. Colonial medicine and public health policies worked hand-in-glove with other political, economic and social mechanisms to restructure Dairen and the rest of southern Manchuria into a colonial space in which diseases and other "natural" manifestations of the local environment and Chinese life would hopefully be replaced by the inoculated, sanitized, and modern world of a Japanese colony.

Modern, Western science in the forms of engineering, urban planning and medicine played a central role in the development of the Japanese colony in southern Manchuria. Just as scientific and engineering technologies embodied in the *Mantetsu* railway, dockyards and soya bean mills provided the transportation and industrial infrastructures necessary for the fledgling Manchurian colony to grow, medical technologies in the forms of hospitals, sanitation facilities, and public health regimens were crucial to the development of the Guandong leasehold. The tremendous financial investment in the region, along with its growing population of Japanese colonial residents, had to be protected from the dangers posed by microscopic enemies. While the Guandong (Kwantung) Army was charged with protecting the region and its railways from both real and constructed security threats, doctors, nurses, and public health officials were also members of an important garrison in the colony. It is important to remember that the modern world that was emerging in the Japanese Home Islands during the Meiji era was also being exported abroad to the growing number of colonies and territories under Tokyo's control.

Notes

Department of History and Classics, Acadia University, Wolfville, Nova Scotia, Canada, B4P 2R6. The author would like to acknowledge the support of the Hannah Institute for the History of Medicine's Grant-in-Aid programme that provided funding for this research.

1. For an introduction to the history of Japanese imperialism and colonial adventures during the Meiji and early Taisho periods see Marius B. Jansen, "Japanese Imperialism: Late Meiji Perspectives," in *The Japanese Colonial Empire, 1895–1945*, ed. Ramon H. Myers and Mark R. Peattie (Princeton: Princeton University Press, 1984), pp. 61–79; Mark R. Peattie, "Japanese Attitudes Toward Colonialism, 1895–1945," in *The Japanese Colonial Empire, 1895–1945*, ed. Myers and Peattie, pp. 80–127; and William G. Beasley, *Japanese Imperialism 1894–1945* (Oxford: Clarendon Press, 1991).
2. On the debates over the concepts of "modernity," "East Asian modernity," and "colonial modernity" (and many other forms of "modernity") and their application to an analysis of the histories of post-Meiji Japan and its colonies see several of the essays in both Gi-Wook Shin and Michael Robinson, eds., *Colonial Modernity in Korea* (Cambridge, Mass.: Harvard University Asia Center, 1999); and Sharon A. Minichiello, ed., *Japan's Competing Modernities: Issues in Culture and Democracy, 1900–1930* (Honolulu: University of Hawai'i Press, 1998).
3. Throughout this chapter I have used the Japanese name for Dairen instead of the city's Chinese name of Dalian. Both names come from the same *kanji* or Chinese characters, and can be translated as "big, or great, connections." I have purposely decided to use the Japanese pronunciation of the port's name as this chapter explores the meanings of "modernity," "development," and "public health" through the historical lens of Japanese visions of their colonial adventure in southern Manchuria. For the same reason I have used the term "Manchuria" to denote the region of northeast China comprised of the Republican provinces of Heilongjiang, Jilin, and Fengtian. In the People's Republic of China, this region is now referred to as *Dongbei*, the "Northeast," in order to avoid using the more "colonially-loaded" term, Manchuria.
4. "Grand Medical Conference at Dairen," *The Manchuria Daily News: Monthly Supplement* (hereafter cited as *MDNMS*), June 1, 1927, 1–3; and *Dairen Iin shinchiku rakusei kinen igaku kaishi* (Medical conference addresses in commemoration of the opening of the new Dairen Hospital) (Dairen: Minami Manshū Tetsudō kabushiki kaisha, 1927), pp. 5–6.
5. "Grand Medical Conference at Dairen," *MDNMS* June 1, 1927, 1–3.

6. Following the 1927 conference all of the prepared research papers were included in a commemorative volume that was published by the SMR. During the conference visiting senior researchers and professors from the Faculties of Medicine at both the prestigious Kyoto and Tokyo Imperial Universities, as well as those from the Manchurian Medical College in Mukden (Shenyang) presented papers, in addition to dozens of local researchers and physicians. Copies of the scientific papers, along with photographs taken during the conference and excerpts of the plenary addresses can be found in *Dairen Iin shinchiku rakusei kinen igaku kaishi* (Medical conference addresses in commemoration of the opening of the new Dairen Hospital) (Dairen: Minami Manshū Tetsudō kabushiki kaisha, 1927). The papers by Drs. Kuno and Suzuki are found on pages 21–28, and 1–32 (in the foreign-language appendix) respectively.
7. "Grand Medical Conference at Dairen," *MDNMS* June 1, 1927, 2; and "Dairen Hospital," *MDNMS* July 1, 1927, 10–13.
8. "Grand Medical Conference at Dairen," *MDNMS* June 1, 1927, 2.
9. On the history of Japan's creation and development of the puppet state of Manzhouguo see Louise Young, *Japan's Total Empire: Manchuria and the Culture of Wartime Imperialism* (Berkeley: University of California Press, 1998); and Prasenjit Duara, *Sovereignty and Authenticity: Manchukuo and the East Asian Modern* (New York: Rowman and Littlefield Publisher, 2004). For a more detailed analysis of Dairen's pre-1931 history see Robert Perrins, "Great Connections: The Creation of a City, Dalian, 1905–1931," unpublished Ph.D. diss., Department of History, York University, Toronto, 1998. The most complete histories of Dairen during the period of Japanese rule are: Gu Mingyi et al., *Riben qinzhan Luda sishinian shi* (A 40-year history of Japan's occupation of Lushun and Dalian) (hereafter cited as *RQLSS*) (Shenyang: Liaoning renmin chubanshe, 1991); and Inoue Kenzaburō, ed., *Dairen-shi shi* (The history of the city of Dairen) (Dairen: Dairen-shi yakusho, 1936).
10. *Minami Manshū tetsudō ryokō annai* (A Travel Guide to the South Manchuria Railway) (Dairen: Mantetsu, 1920 and 1925 editions), pp. 9–12 and 19–20 respectively.
11. For a comprehensive review of the departments and equipment housed in the new SMR hospital in Dairen see *Report on Progress in Manchuria* (hereafter cited as *ROP*) (Dairen: SMR, 1929 ed.), pp. 164–167; "Dairen Hospital," *MDNMS* July 1, 1927, 10–13; and Sun Chengdai and Xu Yuanchen, *Diguozhuyi qinlue Dalian shi congshu weisheng juan* (A collection of materials on imperialist aggression and Dalian's history: Health issues) (hereafter cited as *DSCW*) (Dalian: Dalian chubanshe, 1999).
12. The classic study of the relationship between technology and the creation and development of colonial empires is Daniel Headrick, *The Tools of Empire: Technology and European Imperialism in the Nineteenth Century* (New York: Oxford University Press, 1981).
13. In his analysis of the history of the colonial medical system in the Japanese colony of Taiwan (1895–1945), Liu Shi-yung notes that the terminology used during the colonial period, *shakai eisei* (social hygiene) and *koshū eisei* (public hygiene) were not quite the same as the modern concept of "public health" (*kokuō eisei*). While both the concepts of "public health" and the older "social hygiene" relied on improvements in the fields of epidemiology and bacteriology, the latter also tended to emphasize "social problems and behavior patterns that [were believed to] cause medical crises . . . . In short, the supporters of public health [dealt] with

bacteria, but the scholars of social hygiene focused on people who [were] at high risk or vulnerable to certain diseases. Although both concepts [emphasized] medical progress and prevention, public health usually [changed] the bio-environment to reach its goals, and the advocates of social hygiene [tried] to control people's lifestyle and social behaviors to advance their aims" (p. 8). Both approaches were in evidence during the colonial period in southern Manchuria. See Liu Shi-yung, "Medical Reform in Colonial Taiwan," unpublished Ph.D. diss., Department of History, University of Pittsburgh, 2000, pp. 6–9.

14. For an introduction to the scholarship on the history of medicine in the European colonies in Asia see David Arnold, ed. *Imperial Medicine and Indigenous Societies* (Manchester: Manchester University Press, 1988); Roy Macleod and Milton Lewis, eds., *Disease, Medicine, and Empire: Perspectives on Western Medicine and the Experience of European Expansion* (London: Routledge Kegan Paul, 1988); Mark Harrison, *Public Health in British India: Anglo-Indian Preventive Medicine, 1859–1914* (Cambridge: Cambridge University Press, 1994); and Lenore Manderson, *Sickness and the State: Health and Illness in Colonial Malaya, 1870–1940* (Cambridge: Cambridge University Press, 1996).
15. See Liu, "Medical Reform in Colonial Taiwan" (2000); and Ming-cheng Lo, *Doctors within Borders: Profession, Ethnicity, and Modernity in Colonial Taiwan* (Berkeley: University of California Press, 2002).
16. *DSCW*, pp. 65–68.
17. *DSCW*, pp. 285–290.
18. On the history of the various Guandong administrations see *Kantō-kyoku shisei sanjūnen shi* (A 30-year history of the Guandong Bureau) (hereafter cited as *KKSSS*) (Tokyo: Toppan insatsu kabushiki kaisha, 1936), pp. 61–78; *RQLSS*, pp. 39–42 and 68–77; and Kwantung (Guandong) Government, *The Kwantung Government: Its Functions and Works* (hereafter cited as *KGFW*) (Dairen: Manchuria Daily News, 1929 ed.), pp. 16–21.
19. *DSCW*, pp. 45–50 and 60–63.
20. Great Britain, Foreign Office and Board of Trade, *Diplomatic and Consular Reports: Japan, Dairen* (London: H.M. Stationery Office, 1907, Report no. 3857), p. 8.
21. Koshizawa Akira, *Shokuminchi Manshū no toshi keikaku* (Urban development in colonial Manchuria) (Tokyo: Ajia Keizai Kenkyūjo, 1978), pp. 50–53; and Y. Konishio, *Port of Dairen* (Dairen: Research Office of the South Manchuria Railway Company, 1923), p. 4.
22. See Kantō-shū Chō Chobokuka (Department of Public Works, Government of the Guandong Territory), comp., *Dairen toshi kekaku gaiyō* (A summary of Dairen's city planning) (Dairen: Dairen-shi yakusho, 1938), pp. 1–4; and Liu Zhongquan and Gui Qingxi, eds., *Guanyu Dalian weilai chengshi xingtai de yanjiu* (A study regarding the future of Dalian's urban morphology) (Dalian: Dalian shi ruan kexue keti, 1996), pp. 34–40.
23. The original Russian name for the port, Dal'nii, literally translates as "far away," in reference to its distance from St. Petersburg and the Tsar's government. In the eyes of the Russian railway tsar, Finance Minister Sergei Witte, Dal'nii was to have been developed into the "jewel of the Russian Far East."
24. Konishio, *Port of Dairen*, pp. 3–4; and *Sha E qinzhan Luda de qinian* (Tsarist Russia's seven-year occupation of Lushun and Dalian) (Beijing: Zhonghua Shuju, 1978), pp. 10–12.
25. Adachi Kinnosuke, *Manchuria: A Survey* (New York: Robert M. McBride & Co., 1925), p. 72.

26. On the Russian era in Manchuria the reader is referred to Rosemary Quested, *Matey Imperialist? The Tsarist Russians in Manchuria, 1895–1917* (Hong Kong: University of Hong Kong, 1982); David Wolff, *To the Harbin Station. The Liberal Alternative in Russian Manchuria, 1898–1914* (Stanford: Stanford University Press, 1999); and Soren Clausen and Stig Thogersen, *The Making of a Chinese City: History and Historiography in Harbin* (Armonk: M.E. Sharpe, 1995), pp. 23–52.
27. On the history of the "Garden City" movement in Japan during the early twentieth century see Watanabe Shun-ichi, "Nihonteki Denen toshi ron no kenkyū II: Naimushō chihō kyoku yūshi: Denen toshi (1907) o megutte" ("Studies of the 'Garden City' Japanese style no. 2: An analysis of the introduction of the 'Garden City' concept to Japan in 1907 by the Minister of the Interior"), *Nihon Toshi Keikaku Gakkai gakujutsu kenkyū happyōkai ronbunshū* (The Journal of the City Planning Institute of Japan) 1978, *13*: 283–288; and André Sorensen, *The Making of Urban Japan: Cities and Planning from Edo to the Twenty-First Century* (New York: Routledge, 2002), pp. 89 and 137–142.
28. Kuratsuka's preamble as quoted in Koshizawa, *Shokuminchi Manshū no toshi keikaku*, p. 56.
29. Koshizawa, *Shokuminchi Manshū no toshi keikaku*, pp. 58–59. In particular, Koshizawa notes the residency regulations of *kuri*(s) (coolies) that were detailed in the "Regulations on the Establishment of Special Districts in Dairen," that were enacted by the Guandong Government-General in early 1906. These regulations explicitly stated that: "The residency of *kuri*, as well as other lower-class Chinese, among Japanese in general is not desirable in terms of hygiene and discipline" (p. 59).
30. For fuller discussions of the urban planning of Dairen during the period of Japanese rule see *Dairen toshi kekaku gaiyō* (1938); *RQLSS*, pp. 426–434; and Koshizawa, *Shokuminchi Manshū no toshi kekakum*, pp. 49–57.
31. *Far Eastern Review* (hereafter cited as *FER*), August 1914, 85.
32. On the history of the plague in China see Carol Benedict, *Bubonic Plague in Nineteenth-Century China* (Stanford: Stanford University Press, 1996), pp. 1–15; and Iijima Wataru, *Pesuto to kindai Chūgoku* (Plague and modern China) (Tokyo: Kenbun Shuppan, 2000).
33. Carl F. Nathan, *Plague Prevention and Politics in Manchuria, 1910–1931* (Cambridge: Harvard University Press, 1967), pp. 1–41.
34. Kantō totokufu rinji bōrekibu (Guandong Temporary Municipal Sanitation Bureau), *Meiji yonjū-sen yonen minami Manshū, pesto ryukō shi furoku* (An account of the plague in Southern Manchuria, 1910–1911) (Dairen: Manshū hibi shinbunsha, 1912); and Kimura Ryoji, *Dairen monogatari* (An account of Dairen) (Tokyo: Kenkōsha, 1983), pp. 26–28.
35. See *Meiji yonjū-sen yonen minami Manshū, pesto ryukō shi furoku*, pp. 1–10; and *ROP* (1929 ed.), pp. 171–172. This general attitude was still prevalent when the next outbreak of plague struck Manchuria in 1920–1921. In a report summarizing this struggle against the plague issued by *Mantetsu*'s sanitary office, the SMR's medical superintendent, Dr. Tsurumi wrote: "Hitherto, the low class people, especially Chinese coolies, were regarded as the most dangerous medium. Therefore, these people had to be dealt with adequately, first of all." See *Plague Prevention Campaign in South Manchuria, 1921* (Dairen: Sanitary Office of the SMR, 1921), p. 3.
36. *Meiji yonjū-sen yonen minami Manshū, pesto ryukō shi furoku*, pp. 4–20; and *ROP* (1929 ed.), p. 172.

37. See Richard P. Strong, Erich Martini, G. F. Petrie, and A. Stanley, eds., *Report of the International Plague Conference Held at Mukden, April 1911* (Manila: Bureau of Printing, 1912), pp. 33–34; Kantō-kyoku (The Guandong Bureau), *Kantō-kyoku tōkei sanjūnen shi* (Thirty years of statistical records of the Guandong administration) (Dairen: Kantō-kyoku, 1935), pp. 648–651.
38. For more detailed accounts of the plague outbreak in Manchuria during the early twentieth century see Nathan, *Plague Prevention and Politics in Manchuria, 1910–1931* (1967); *ROP* (1929 ed.), pp. 171–177; *DSCW*, pp. 48–56; Wolff, *To the Harbin Station*, pp. 92–95; and the autobiography of Wu Liande, the Chinese physician who headed the North Manchurian Plague Prevention Service for much of the early twentieth century, Wu Lien-teh, *Plague Fighter: Autobiography of a Chinese Physician* (Cambridge: W. Heffer and Sons, 1959).
39. On the development of the soya bean trade and the construction of oil mills in Dairen during the years of World War I see FER, October 1925, pp. 666–667; and *RQLSS* pp. 304–308.
40. On Dairen's growth as a port during the soya bean boom see Gu Mingyi et al., *Dalian jin bai nian shi* (The history of Dalian's last hundred years) (hereafter cited as *DLJBNS*) (Shenyang: Liaoning renmin chubanshe, 1999), vol. 1, pp. 842–854; *RQLSS*, pp. 234–260; and *China Economic Monthly*, August 1924, 24.
41. See *Kantō-chōkan kambō bunshoka* (Archives Department of the Secretariat of the Guandong Governor), *Kantō-chō tōkei nijūnen shi* (Twenty years of statistical records of the Guandong administration) (hereafter cited as KCTNS) (Dairen: Manshū nichinichi shinbun, 1927), pp. 13–15. In 1914 the total population of Dairen was 121,933, of which 38,436 were Japanese, 83,396 were Chinese, and 101 were "others." By 1920 the port city's population had grown to 238,867, of which 62,994 were Japanese, 175,721 were Chinese, and 152 classified as "others."
42. *KGFW* (1929 ed.), p. 72; *FER* June 1922, 376; and *RQLSS*, pp. 426–427.
43. *KGFW* (1929 ed.), pp. 72–73 and 85–86; and *FER* September 1925, 592–593.
44. In his examination of the history of urban planning in the Japanese colony in Manchuria, Koshizawa Akira points out that the zoning system developed in Dairen (residential, commercial, industrial, and combined) was not adopted in Japan until the late 1920s; another example of how developments in the south Manchurian colony were sometimes in advance of, or more "modern" than was the case in Japan proper. See Koshizawa, *Shokuminchi Manshū no toshi keikaku*, p. 53. On the Japanese attitudes toward disease and nationality in Dairen see *DLJBNS*, vol. 2, pp. 1515–1517; and Koshizawa, *Shokuminchi Manshū no toshi keikaku*, pp. 49–50 and 58–60.
45. *KGFW* (1929 ed.), p. 72.
46. "Annual Report for the Year 1907, Sir Claude MacDonald, Britain's Ambassador to Japan," FO 881/9218(1), in *British Documents on Foreign Affairs: Reports and Papers from the Foreign Office Confidential Print. Part 1. Series E. Asia, 1860–1914. Vol. 9. Annual Reports of Japan, 1906–1913*, ed. Ian Nish (Washington, D.C.: University Publications of America, 1989), p. 62. Of the 30,000 Japanese who arrived in the spring and summer of 1906, more than 16,000 left the following year, having failed to find the proverbial pot of gold in southern Manchuria.
47. *North China Herald and Supreme Court and Consular Gazette* (Shanghai), September 7, 1906.
48. *KCTNS*, p. 8.

49. On the formation and responsibilities assigned to the Marine Bureau office in Dairen see *KKSSS*, pp. 1003–1009; *KGFW* (1934 ed.), pp. 47–49; and *DSCW*, pp. 232–237.
50. See *KGFW* (1934 ed.), p. 93; *ROP* (1929 ed.), pp. 163–164; and *DSCW*, pp. 232–241.
51. *DSCW*, p. 83 and pp. 231–236.
52. See Manderson, *Sickness and the State*, p. 101; and chapter 4 in this work titled, "Public Health and the Pathogenic City," pp. 96–126.
53. Manderson, *Sickness and the State*, p. 97.
54. *DSCW*, pp. 205–227; *KGFW* (1929 ed.), pp. 40–43; M. Tsurumi, "Public Hygiene in Manchuria and Mongolia," *The Light of Manchuria*, February 1, 1921, pp. 1–4; and "Hygiene of South Manchuria," *The Light of Manchuria* August 1, 1922, pp. 5–6 and 25–47.
55. *Manchuria Daily News* (hereafter cited as *MDN*) April 2, 1915; and *KGFW* (1929 ed.), pp. 40–41.
56. "Hygiene of South Manchuria," *The Light of Manchuria* August 1, 1922, 5 and 25–26; and *ROP* (1929 ed.), pp. 163–167.
57. *FER*, August 1914, p. 85. For more on the importance of the SMR to the development of the prewar Japanese colony in Manchuria see Yoshihisa Tak Matsusaka, *The Making of Japanese Manchuria, 1904–1932* (Cambridge, Mass.: Harvard University Asia Center, 2001).
58. For examples of the *Mantetsu* and colonial administration's descriptions of Dairen as a model city see *Minami Manshū tetsudō ryokō annai* (1920 and 1925 editions), pp. 9–12 and 19–20 respectively; and *Dairen chihō annai* (Guide to the Dairen area) (Dairen: *Mantetsu*, n.d.). More recent nostalgic Japanese publications on the history of Dairen continue to use terms such as "city of lights," and "beautiful port city." For examples of this sort see Kimura, *Dairen monogatari*, pp. 35–37; and Suzuki, *Jitsuroku Dairen kaisō*, pp. 1–4 and 76–84.
59. Manderson, *Sickness and the State*, p. 101.
60. *ROP* (1929 edition), p. 163; and M. Tsurumi, "Public Hygiene in Manchuria and Mongolia," *Light of Manchuria*, February 1, 1921, 3–4 and 10–12.
61. See *KCTNS*, pp. 232–237. The year 1917 is representative of the general numbers of fatalities from dysentery and intestinal illnesses, with the recorded total deaths for each category standing at 137 and 3,308 respectively.
62. See *KCTNS*, pp. 232–237.
63. See William Johnston, *The Modern Epidemic: A History of Tuberculosis in Japan* (Cambridge, Mass.: Council of East Asia Studies, Harvard University, 1995), pp. 70–90.
64. *KCTNS*, pp. 232–237. Again, the figures for 1917 are fairly representative of the period, with 281 Japanese, and 1,104 Chinese succumbing to the disease.
65. "Hygiene of South Manchuria," *The Light of Manchuria*, August 1, 1922, 40–41; *DSCW*, pp. 99–125; and Kinoshita Suzuo, *Dairen Seiai Iin Nijūgoshūnen shi* (A 25-year history of the Dairen Seiai Hospital) (Dairen: n.p., 1931).
66. *DSCW*, pp. 134–138.
67. *MDN* October 25 and 26, 1918.
68. A few of the standard works on the history of the 1918–1919 influenza pandemic are Richard Collier, *The Plague of the Spanish Lady: The Influenza Pandemic of 1918–1919* (London: Macmillan, 1974); Alfred W. Crosby, *Epidemic and Peace, 1918–1919* (Westport: Greenwood, 1976); and Howard Phillips and David

bars and reinforced concrete in almost all of their projects in Tokyo, while many Japanese construction firms had not, resulting in a clear lesson when the Kantō earthquake struck on September 1, 1923.

97. *FER* January 1924, p. 31; and *DSCW*, pp. 99–101.
98. *MDNMS* June 1, 1927, p. 1.
99. For detailed descriptions of the hospital's layout and facilities see: *MDNMS*, July 1, 1927, pp. 12–13; August 1, 1927, pp. 8–12; and *DLJBNS*, Vol. 2, p. 1465.
100. "SMR Hospitals: What Sacrifice is Paid for Their Maintenance," *MDNMS*, July 1, 1923, p. 15.
101. *MDNMS* November 1, 1927, pp. 12–13.
102. As late as 1929, three years after the SMR's showcase medical facility opened, average weekly wages in Dairen were still far below the daily room charges at the hospital. Weekly wages for skilled Japanese laborers ranged from four and a half yen for a mason at the upper end, to three yen for a printer at the bottom end of the range. Chinese wages were roughly half of those paid to a Japanese worker similarly employed. See *ROP* (1931 ed.), pp. 173–174.
103. *DSCW*, pp. 109–117.
104. R. O. Matheson, *Modern Manchuria: A Series of Articles Written for the Chicago Tribune* (Dairen: Manshū nichi-nichi shinbun, 1926), p. 8.

Part 2

Technology, Industry, and Nation

6

The Mechanization of Japan's Silk Industry and the Quest for Progress and Civilization, 1870–1880

David G. Wittner

Introduction

In the decade following the Meiji Restoration, Japan embarked on a far-reaching program of industrialization the likes of which the world had never seen before, nor is it ever likely to see again. The Meiji government's "program of industrialization," *shokusan kōgyō*, may, however, be more accurately described as *ad hoc* industrialization: a series of perfunctory ventures whose only elements of commonality were the adoption of Western industrial technologies that loosely fit within the rhetoric of *fukoku kyōhei* ideology permissible under the unequal treaties. There was little or no detailed planning involved; schemes were often formulated as problems arose. At Tomioka, the government's premier silk reeling facility, for example, no one even considered who would work there.[1] This lack of planning and foresight was typical of early Japanese efforts at technology transfer and industrial development.

From the perspective of technology transfer, the first decades of industrialization can be roughly divided into two periods: from 1868 to approximately 1884, and from approximately 1884 until 1895. The dividing line between these periods is the publication of *Kōgyō iken*, a government report that was intended to solve the problem of "trial and error industrialization." The project of Maeda Masana, a *Nōshōmushō* (Ministry of Agriculture and Commerce) bureaucrat, *Kōgyō iken*, was a countrywide survey of the actual conditions of Japan's industries. In it, Maeda attempted to present a comprehensive appraisal of Meiji economic policy. Among his suggestions, Maeda urged that the government reconsider the direct transfer of imported technology and give preference to the modernization of rural industry.[2]

Within the first decade of Meiji industrialization the Iwakura mission, the Meiji government's first official tour of Western Europe and the United States, provided for subtle changes to the ways technologies were selected.

Prior to the mission, the government's choice of technique was based on the decisions of a core group of officials with little or no technical expertise. They relied on personal connections—including associations continued from *bakufu*- or *han*-based business ventures, Western officials' recommendations reflecting personal ambition, and most importantly the presumption that importing Western technology would bring "civilization" to Japan. Pre-mission technology transfer must be viewed as part of the new government's effort to bolster its position vis-à-vis the *han* and defunct *bakufu* and as an attempt to present Japan as "civilized" in the Western world order.

Post-mission choice of technique is more indicative of the Meiji government's analysis of a country's political and/or economic standing, although this too, had a role in the selection of technologies in the pre-mission years.[3] In both periods, technologies were frequently imported based on their presumed abilities to bring Western "civilization" to Japan, regardless of technical rationality.[4] From the government's perspective, importing what it believed to be the most modern methods and the most modern machinery—made from the most modern materials—would be indicative of the extent of "progress and civilization" in Japan.

In its efforts to renegotiate the unequal treaties, to build a "Rich Nation and Strong Army," the Meiji government's "program of industrialization" was based on having Japan conform to Western ideals of progress and "civilization" in developed countries. It was not necessarily based on a technical examination of conditions in Japan that would facilitate or hinder the transfer of a specific technology. Moreover, the government's insistence on importing only what it perceived as the most advanced technologies was, at times, counterproductive to its stated goals.

Through an examination of government-led initiatives and private efforts to mechanize Japan's silk reeling industry in the decade following 1868, this chapter will demonstrate that the preeminent consideration that guided choice of technique and technology transfer was more ideological than technical or economic. In government sponsored enterprises, beliefs in "modernity" and material representations of authority, progress, and "civilization" were more important for choice of technique than any technical assessment of a technology's appropriateness.

Mechanization of the Silk Industry: Selecting a Foreign Adviser

In late 1869, Vice Minister of Finance Itō Hirobumi was approached by F. Geisenheimer, the manager of a French trading company in Yokohama, who complained about the declining quality of Japanese silk. As a remedy, he recommended establishing a model filature where Japanese silk producers could learn Western reeling techniques to the commercial (and technological benefit) of all involved. More than completely overhauling Japan's domestic reeling industry, however, Geisenheimer was more concerned with profit. Itō demurred, citing possible treaty violations as the reason. Insistently, however,

Geisenheimer suggested a Japanese run facility that relied on French capital and technology imports, but once again, Itō declined. He was intrigued, however, by Geisenheimer's insistence as to the profitability of a model facility based on Western technology. Shortly thereafter, Itō initiated discussions between the *Minbushō* (Ministry of Civil Affairs) and *Ōkurashō* (Ministry of Finance) regarding the feasibility of such a project. After some deliberations, officials within the two ministries decided to hire foreign advisers and proceed with the project. Itō contacted Shibusawa Eiichi, head of the *Ōkurashō*'s Taxation Bureau and the only government official with any silk-related experience, to help find a suitable Western adviser.[5] In February 1870, Itō and Shibusawa went to Tsukiji, Tokyo where they approached Lieutenant Albert Charles Du Bousquet to help the government find a silk reeling expert. On the recommendation of Ōkuma Shigenobu, they also went to Yokohama where they again spoke with Geisenheimer about the venture. Geisenheimer was the branch manager of the Lyons-based silk wholesaler Hècht, Lilienthal and Company. Du Bousquet and Geisenheimer recommended that the government hire Paul Brunat, a young Frenchman who had been working for Hècht, Lilienthal and Company in Yokohama for about two years.[6] Brunat had been sent to Yokohama by the Lyons silk wholesaler, where he had worked in the silk industry since he was a teenager.[7] The four men visited Brunat in June 1870, at which point Itō and Shibusawa decided to hire him as the government's foreign silk reeling adviser. Shortly thereafter, Brunat was given a provisional contract. His final contract was signed in November of the same year.[8]

This rather circuitous route by which the Meiji government found and hired its primary silk reeling adviser is illustrative of pre-Iwakura mission methods of securing technical advice. Understandably, the urgency of Japan's political situation, international and domestic, meant that formulating a carefully delineated set of hiring guidelines would take the back burner to more pressing issues of state. As Hazel Jones has noted, "time was the missing quantum; day-to-day, hour-to-hour decisions in this period of revolutionary change precluded thoughtful consideration of long-term effects."[9] When Brunat was hired, trust was the primary consideration upon which the hiring of a foreign adviser was based. While theoretically including an assessment of a potential adviser's qualifications, this practice also appears to have caused its own set of problems. The policy statement, which guided the hiring of foreign advisers, *Gaikokujin yatoi irekata kokoroe jōjō* (Instructions for the Hiring of Foreign Employees), issued in February 1870, was continuously modified over the next 20 years in an effort to better regulate the engagement of foreigners.[10]

Profit and a steady supply of high quality raw silk were the prime motivating factors in Geisenheimer's and Du Bousquet's recommendation of Brunat. Indeed, Brunat was continually employed by Hècht, Lilienthal and Company until 1873—three years after signing his contract with the Meiji government.[11] The French community in Yokohama also shared this view, reacting with enthusiasm and the natural assumption that the Lyons connections would translate into equipment purchases from France.[12] In fact, according to the terms of Brunat's final contract, all the equipment for Tomioka was imported through Hècht, Lilienthal and Company and Geisenheimer was one of the signatories.[13] The Meiji government, however, may have had other ideas about this new relationship; for Meiji Japan's "special relationship" with France was also political and ideological.

It could be argued that at the time finding foreign advisers through personal connections was the only option available to the Meiji government. One could also stress that connections were, and remain to this day, important in Japan. However, the government's mixed record of success and official attempts to remedy the situation does not support these inferences. Part of the problem was the nature of the early Meiji government itself. As Umegaki Michio has demonstrated, the new government suffered from any number of organizational deficiencies. Prior to late 1871, there were no firmly established procedures for examining and screening policy proposals or for reviewing their development and outcome once implemented. Shibusawa Eiichi vehemently complained about the lack of organization and direction in the central bureaucracy and was instrumental in having Ōkuma Shigenobu organize a committee that would eventually be responsible for implementing policy procedure.[14] Until that time, however, appointments to central positions were made through closed-door negotiations and conversations among a few officials.[15] The hiring of foreign advisers was haphazard: "As a task arose an immediate solution was sought, and someone hired."[16] The *ad hoc* and individualized nature of the early Meiji bureaucracy rendered the adviser selection process risky at best. It was perhaps this realization, coupled with the new government's demands for importing "civilization" to Japan, that often led government officials to ignore their adviser's recommendations.

Choice of Technique: A Technological Assessment

If the method by which the Meiji government secured its foreign advisers was somewhat haphazard, the criteria on which officials based their choice of technique appears even less purposive. Brunat recommended that the government build a sizable facility that would utilize a Japanese–European hybrid reeling technology. Because the Tomioka area was already famous for its high quality *zaguri* reeling techniques, and because he did not want to disturb local production patterns, Brunat suggested surveying local reeling methods and incorporating them into Western mechanized reeling processes. He recommended locally produced machines, made of materials that were available in Japan, that is, wood, and also suggested a local survey so as to ascertain what changes to present (Japanese) reeling methods would be most beneficial. He also sought to find out how resistant local silk reelers would be toward changes in their industry. From the perspective of choice of technique, Brunat's proposal was relatively conservative. Although he recommended the incorporation of steam reeling techniques, steam engines to power the mill, and proposed a large facility—300 basins—much of his plan was based on utilizing local technology.[17]

Government reactions, while seemingly enthusiastic, must have been a source of frustration for Brunat. On November 29, 1870, the day he received his final contract, he was ordered to return to France to purchase everything required literally to import a fully operational French filature. Brunat was ordered to buy the latest (cast iron) reeling frames, steam engines, machine tools, and other equipment.[18] The government's directive called for a facility, the design of which was to be based primarily on the government's specifications.[19] Meiji leaders' desire to use the latest Western technologies bordered on obsession. The issue fuel for the steam engines and boilers illustrates this especially well: Brunat recommended wood over coal because of the latter's exorbitant price. Meiji bureaucrats, however, opted for coal and not just any coal. It too, was to be imported from France![20] In the end, Tomioka would be based strictly on orthodox French reeling technology.

From what evidence remains, it appears that Brunat's proposal was rejected through the decisions of a few men with little or no experience in the silk industry. Of the five officials directly responsible for Tomioka, one, Shibusawa Eiichi, had experience in sericulture. Odaka Atsutada, Shibusawa's cousin, brother-in-law, and ultimately the facility's day-to-day supervisor, was from a merchant family that dealt in indigo. The remaining three officials, Tamanoe Seiri, Nakamura Michita, and Sugiura Yuzuru were former samurai with little or no commercial or technical experience.[21] Two additional men, Ōki Takatō and Yoshii Tomozane, the officials of record in the *Minbushō*, were also former samurai with apparently no technical or commercial background.

It is often pointed out that many Meiji bureaucrats studied overseas, if only briefly, and that this was an asset in their development of an industrialization policy. Itō Hirobumi, Inoue Kaoru, and Yamao Yōzō studied in England for six months in 1863; Shibusawa and Sugiura similarly studied in France in 1867. But how much they actually learned is questionable. Shibusawa himself admitted only to being capable of the most rudimentary French, making tasks such as shopping possible without an interpreter.[22] At issue, however, should not be whether these men had technical backgrounds, but that they chose to ignore the advice for which they were paying in the form of foreign advisers. For Meiji policymakers committed to industrialization, overseas experience legitimized cultural materiality for *bunmei kaikai* ideology. In sum, the government's actions with respect to technique/technology clearly indicate an ideological alternative agenda.

Although the official record presents no clear evidence as to the selection process, there are indications that Shibusawa Eiichi and Sugiura Yuzuru were behind the decision to adopt French reeling technology. In 1867, the two were part of a Tokugawa mission to Europe where they had the opportunity to visit the Fifth International Exhibition in Paris.[23] Ultimately Shibusawa would live in Paris for six months serving as an attendant for Tokugawa

Akitake, the Shogun's younger brother.[24] Traveling to Lyons by train, Shibusawa and Sugiura had only a brief opportunity for sightseeing but were impressed nonetheless. They noted the spaciousness and metropolitan character of Lyons and were particularly keen on the variety and quality of locally produced women's silk accessories. While neither man had the opportunity to visit a reeling or weaving facility while in Lyons, they were able to examine its silk more closely at the Paris exhibition in June. In Paris, both Shibusawa and Sugiura were impressed, not only by the quality of the silk fabrics made in Lyons, but also by the sheer magnitude of the exhibition, not to mention the city itself. Moreover, they noted the prestige attached to winning awards at such an exhibition and how countries vied for distinction by displaying their most advanced and finest products.[25]

After the exhibition ended, the two toured Switzerland, Italy, Great Britain, and the Netherlands. In September, Shibusawa and Sugiura visited filatures and weaving establishments in Switzerland that they described as being merely artisanal or handicraft in nature. While commenting favorably on the detail of some of the fabrics, the two were obviously less impressed with the quality of what they had seen and with the small size of the facilities they had visited. Their travels through Italy, while including a trek through the mountains in a horse-drawn carriage, a far cry from the steam locomotives to which they had become accustomed, did not include a visit to a filature. Silk was not even mentioned with regard to Italy. Shibusawa described the carriage in detail, however, noting that it was the means of transportation used in Europe *before* the invention of the train.[26]

It is likely that this first quick impression of Western weaving and reeling methods, coupled with their awe of Paris and Lyons, biased their opinions in favor of French reeling techniques. By the 1860s, the reputation of Lyons silk was well known in Japan, its quality considered among the best. Moreover, there existed what historian Richard Sims calls a "special relationship" between France and Japan that was largely the result of Leon Roche's efforts, although the relationship predated his 1864 arrival.[27] As early as 1861, the export silk trade with France was significant, totaling some 2,600,000 francs. During Shibusawa's time in Paris, the volume of trade increased by a factor of seven, to 20,220,000 francs; in 1868, shortly after his return, it again more than doubled.[28] The "special relationship" was even deeper than Sims suggests. Sigismond Lilienthal, partner in Hècht, Lilienthal and Company, the Lyons-Yokohama trading firm that employed and recommended Brunat, was instrumental in orchestrating arms-for-silk deals with the new Meiji government.[29]

At the time, the quality of French silk, Lyons silk in particular, was known in Japan; the process by which it was made, however, was not.[30] French silk reeling technology was not, in fact, purely French; local reelers had long since modified their technology with Italian designs. Indeed, by the mid-nineteenth century, Italian reeling methods were widespread throughout France and were considered superior by many silk reelers to traditional French methods, a point that seems to have been lost on these early observers from Japan.[31]

If the government's representatives did not understand the technical details of French and Italian silk reeling machinery, they certainly recognized the political and economic importance of the two countries for Japan. Diplomatic relations with Italy were formally established in 1866; commercial relations in the form of the raw silk and silkworm egg trades through Yokohama were established shortly thereafter.[32] Much for the same reasons that the French were forced to look outside their borders for raw silk and silkworms, the pebríne virus also brought Italian silk merchants to Japan. The disease had much more serious consequences for France's industry, however, and the volume of trade with Italy was correspondingly lower. As noted earlier, not only did Shibusawa's trip to Italy bypass silk reeling facilities, but also, years later the Iwakura mission relegated the industry in general to an unofficial side-trip. Indeed, our knowledge of the mission's silk-related activities comes from the Italian press, not the official record of Japan's visitors in Italy.[33] Although Japanese sericulturists were later sent to Italy to study scientific methods of silkworm rearing, Italy was seen more as a source of competition as was indicated in an 1873 report by the Japanese consul-general to Italy. In it, he notes that the Italian government was licensing trading companies with the purpose of promoting silkworm rearing and eliminating the imports of (Japanese) silkworm egg cards.[34]

More crucial for the new Meiji government, however, was their evaluation of the two countries' political importance. Shibusawa noted in his travel log for October 23 (1867) that Italy had been torn by civil war a year before the Tokugawa mission arrived. He also noted that Tokugawa Akitake's audience with King Victor Emmanuel II was cut short because of disturbances in Rome. In fact, Rome would not officially be part of the unified Kingdom of Italy until 1870. While Shibusawa praised the skill of the artisans at a mosaic factory, these comments pale in comparison to the traveler's evaluations of French manufacturing and Paris.[35] Much like the low-level of American diplomatic participation in Japan following the U.S. Civil War, Italy was similarly relatively inactive in early Meiji politics—especially in comparison to Britain and France. Although Italy made contributions to Japan's military technology during the Meiji period, its greater contributions are in the introduction of Western art and music in the late 1870s.

As a result of France's greater participation in Japanese politics and trade, the impressions of the country's technical and political prestige formed at the Fifth International Exhibition in Paris, and the aura of "civilization" and modernity, which—for Shibusawa and Sugiura—seemed to exude from every feature of French society, the Meiji government decided to import strictly orthodox French silk reeling technology for its model filature at Tomioka. There was no technical evaluation of the machinery, nor was there any attempt to ascertain whether this new technology would be appropriate for Japan or Japanese silk reelers.

Determining choice of technique is a difficult process regardless of timing or location. In Meiji Japan, the task was further complicated by the nature of the government, internal disturbances, and external conditions, such as the unequal treaties and the need to demonstrate the legitimacy of the new Imperial government to the leading Western powers. An issue that adds to the complexity of the debate surrounding the government's decision to import orthodox French technology is that there *were* alternative Western mechanized reeling techniques available in Japan whose technologies were relatively simple to import. Arguably primitive when compared with Tomioka's technology, these alternative reeling frames were capable of producing silk of a more uniform and higher quality than traditional Japanese methods. In the hands of a trained operator, they could reel silk of a quality equal to that of Tomioka. More importantly, they were easily fabricated by local craftsmen using locally available materials.

Materials, "Civilization," and the Technological Alternatives

The chief critic of the government's plan was Hayami Kenzō, the man responsible for first importing Western mechanized reeling technology to Japan, and often considered the father of Japan's modern silk industry. Hayami questioned the feasibility of the facility noting that its (French) technology was too expensive and complicated "to be of any immediate benefit to Japan's silk reeling industry" or to the Meiji government. Based on his assessment of the government's proposal, Hayami also questioned whether or not Brunat was qualified. As McCallion notes, this was the first time anyone seriously questioned the government's plans.[36] This would not be the last time that Hayami leveled an attack on government-led silk mechanization initiatives; and it would also not be the last time the Meiji government chose to ignore the advice of its own advisers.

Hayami was in a unique position from which to criticize. Working originally for Maebashi-*han*, he first established a retail silk outlet in Yokohama in March 1869. For a samurai, Hayami was a rather astute businessman. He surveyed Western merchants in Yokohama and found that the chief complaint with Japanese silk was not the raw material, as had previously been assumed, but the poor quality of Japanese reeling techniques.[37] In much the same way that Shibusawa and Itō were introduced to Brunat, Hayami sought and received an introduction through the Swiss consul in Yokohama to Casper Mueller, a Swiss silk reeling expert, who had been involved in Italy's silk reeling industry for over a dozen years.[38]

Following Hayami's guidance and Mueller's advice, with approval from the new Meiji government, Maebashi-*han* established the first mechanized reeling facility in Japan based on Western technology in June 1870. The filature was small, initially with only three reeling frames. It relied on Italian technology, and all of the machinery was locally produced. Before the Meiji government could even complete construction on Tomioka, the Maebashi filature had moved once and expanded twice. Still based on locally manufactured, wooden, Italian-style machinery, Maebashi filature by June 1871 boasted 36 reeling frames.[39]

Unfortunately, the Maebashi facility was neither very profitable nor long lasting. It was important, however, for providing an alternative technology to that employed at Tomioka. The two facilities were built with one goal in common; to serve as models of mechanization and modernization for Japan's local silk reelers. In this sense, Maebashi was far more practical than Tomioka; its (simpler) Italian-style machinery was more within the grasp of Japanese producers—financially, methodologically, and technologically—than Tomioka's French technology. Moreover, it was the "first attempt to introduce Western reeling techniques as much as possible within an exclusively Japanese context."[40] Whereas neither facilities' technology was replicated with any significant degree of accuracy, private silk reelers were able to copy the Italian technology far more faithfully than the French. In fact, within five months of its opening, the first group of trainees to visit the model filature arrived at Maebashi.[41] From the perspective of aiding the basic transfer of technology to local producers, Maebashi would have to be judged more successful.

The first attempt to replicate the government's model filature came in 1874 when a group composed largely of former samurai opened a filature in Nagano prefecture, Rokkōsha, theoretically based solely on Tomioka's French technology. Yet the fact that Rokkōsha was able to serve as a model for others to follow later, betrays the simplicity of the operation. Rokkōsha did not have the government's financial resources, as was the case with Tomioka, and modifications had to be made. These primarily came in the form of cost-saving adaptations to the technology. All of Rokkōsha's reeling frames were locally produced from locally available materials. The copper basins for which Tomioka was famous were replaced with ceramic. Steam engines were replaced by a waterwheel, and kindling fires replaced coal. Yet, through all the changes, Rokkōsha still boasted that it produced silk by Tomioka's superior French methods.[42]

Tomioka, Maebashi, and Rokkōsha were visited by silk reelers from all parts of Japan who wished to improve their reeling methods. At the very least, they were curious to see what European technology had to offer and how this could be turned into profit. In many cases, these small-scale producers imported some aspect of what they had seen into their facilities, but even the most painstaking attempts at replicating Tomioka's technologies were far from faithful to the original model. More often than not, however, firms advertised their silk as being made in a "Tomioka-style" facility or by "Tomioka-style" methods after adopting only the most insignificant aspects of the technology. Few, if any, claimed to have modernized through the adoption of Maebashi's technology, although its methods were more widely diffused and its silk had a good reputation and drew high prices in international markets.

Even the small-scale producers who had visited Rokkōsha, but who had never visited Tomioka, and who further simplified the technologies to match their particular situations, claimed to produce silk by Tomioka's methods. A Saitama prefecture filature established in 1876 that claimed for example, to be based in part on Tomioka's technology was actually modeled on another

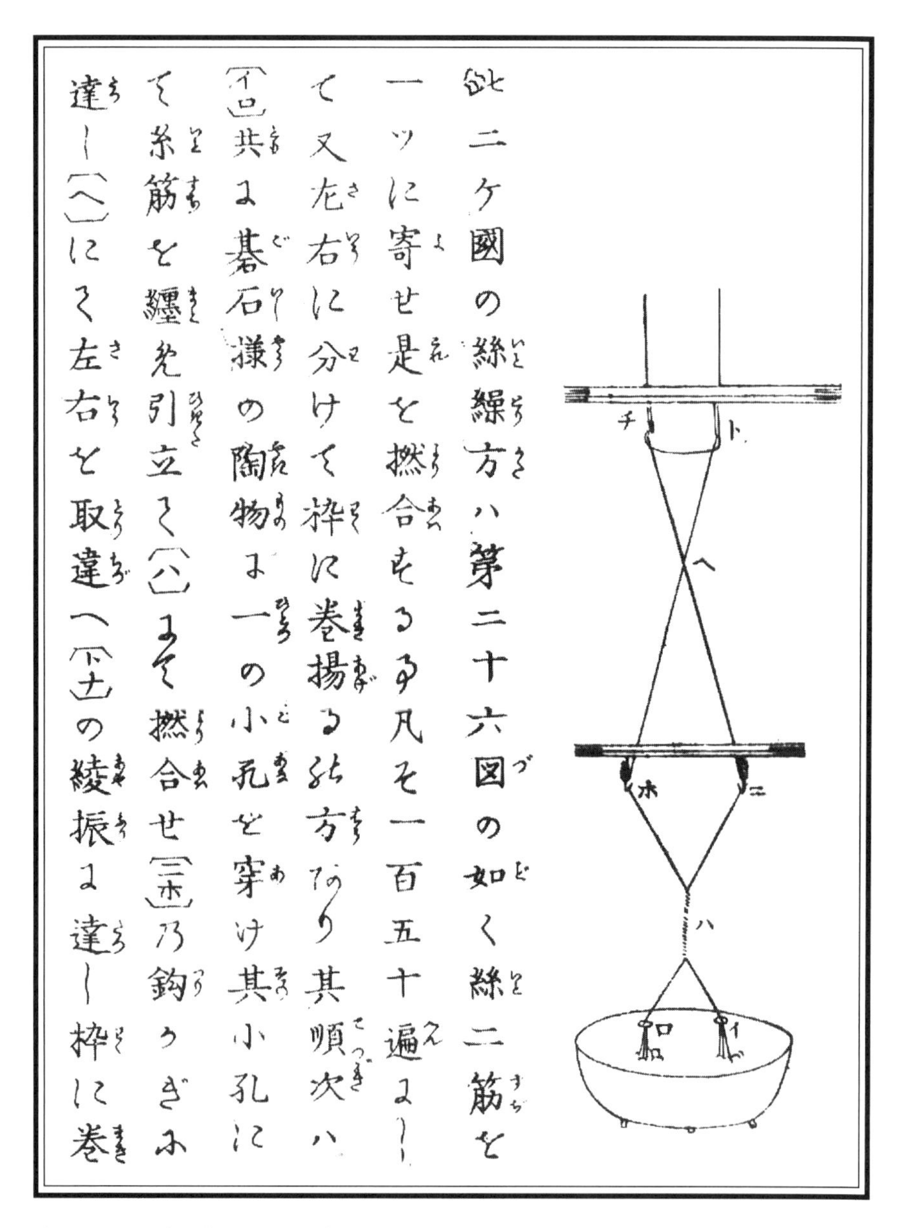

此二ヶ國の絲繰方ハ第二十六図の如く絲二筋を一ツに寄せ是を撚合をするヿ凡そ一百五十遍よりて又左右に分けて枠に巻揚る糸方なり其順次ハ〔イロ〕共に碁石様の陶物に一ツの小孔を穿け其小孔にて糸筋を纏ひ引立て〔ハ〕にて撚合せ〔ニホ〕乃鉤[illegible]ぎふ達し〔ヘ〕にて左右を取違へ〔トチ〕の綾振に達し枠に巻

Figure 6.2 *Chambon* method of croisure. Beginning at the bottom, two sets of four to six silk filaments are gathered together into two threads, which are in turn twisted together and then separated. The intertwining of the two threads strips excess water, cleans, and also binds the individual filaments into a single thread. The two threads are then reeled onto separate take-up reels. This method, adopted at Tomioka, is also called *toyomori*.

Source: Date Saburō, *Kiito seihō shinan* (Tokyo, 1874).

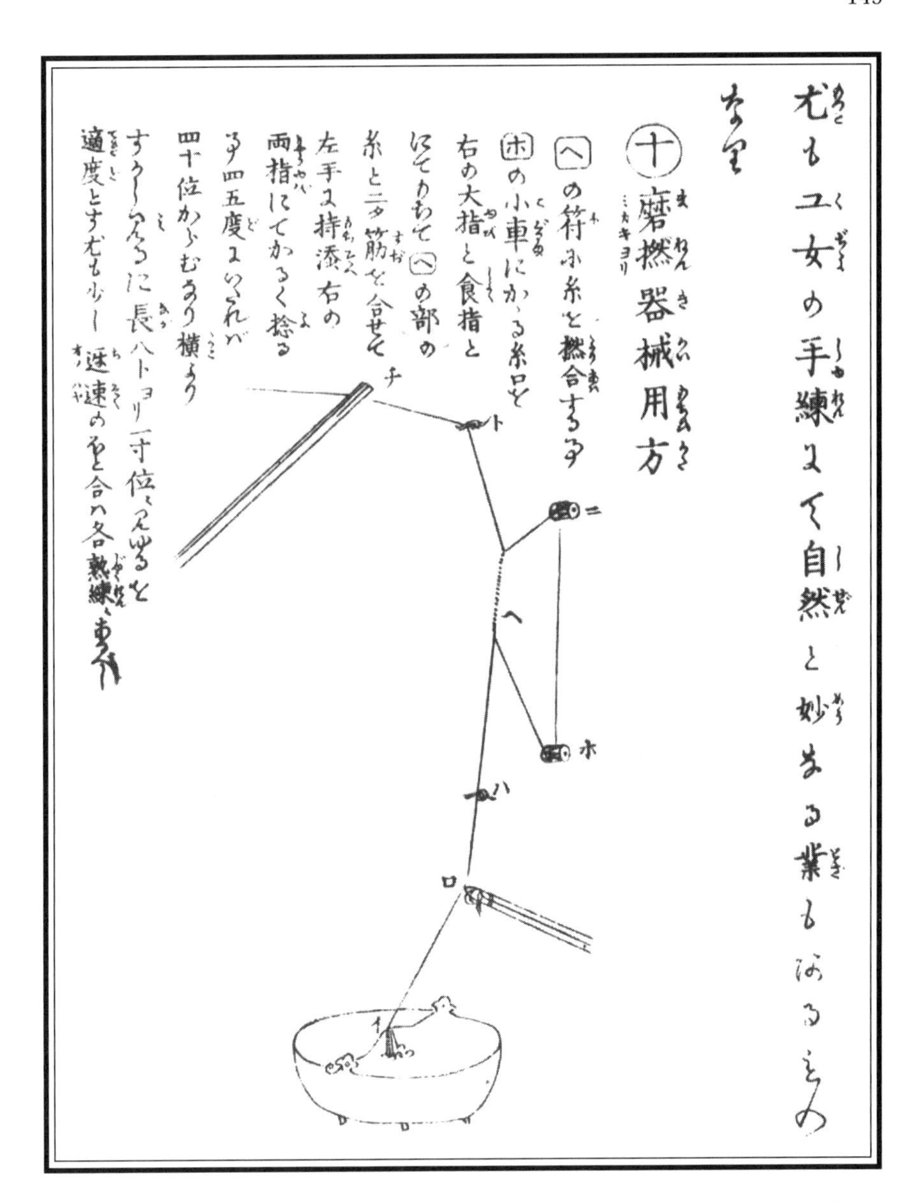

Figure 6.3 This is a *tavelle*. A single thread composed of four to six silk filaments is passed over a number of small wheels and twisted over itself, achieving the same results as with the *chambon*. This method had the distinct advantage of being easily adapted to local technology. In Japanese this method, which was used at Maebashi and Akasaka, is also referred to as *kenneru*.

Source: Date Saburō, *Kiito seihō shinan* (Tokyo, 1874).

Economic Perspectives on *Choice of Technique*

In an effort to better understand the Meiji government's position regarding the technology it chose to import and promote, it may be appropriate to examine briefly factors that are often considered important for choice of technique. The most basic arguments focus on economic factors that are considered determinative for technological choice. Proponents of this position argue that entrepreneurs will adopt the methods that give them the greatest output for the smallest input. Issues, which add to the complexity of this model, are the structure of related technologies within preexisting industries, and the extent of the distribution of complimentary or alternative existing technologies. Given these conditions, for the Meiji government the "proper" choice of technique would have been the one that: (a) provided the most favorable ratio of increased productivity to minimal capital outlay; (b) fit largely within the structure of, and enhanced, the existing silk industry; and (c) could be adopted within the context of the existing technology, that is, adoption of a new reeling technology would not require completely abandoning the previous manufacturing system. As Penelope Francks notes in her study of innovation in Japanese agriculture, however, the aforementioned model does not take local conditions, such as the physical environment and climate, infrastructure, and availability of materials into account; neither does it address such external issues as economic assistance or education dependency.[61]

From this perspective, the Meiji government's adoption of French reeling technology in the form of a turnkey facility should be considered economically irrational. Brunat's original proposal focused on building a filature that incorporated modern techniques, such as steam reeling and steam mechanization, into a context that was technologically and economically appropriate for conditions in Japan. He tried to address the issue of incorporating Western technology into existing methods. He urged that there be no break with tradition and suggested that the government incorporate recent European technological advances that would help raise the quality of Japanese silk and efficiency of the industry in general.

While economic issues figured strongly in Brunat's prospectus, most of his recommendations were rejected. He noted that importing all the necessary machinery was expensive, and repairing cast iron machines would be nearly impossible given the state of Japan's iron industry. The delays involved with obtaining replacement parts from France would be extreme and the costs exorbitant. The French process also was significantly different from local methods. It was more complicated and it would take a long time for inexperienced workers to gain any degree of proficiency. Brunat argued against any abrupt changes to the industry that would be inefficient or expensive. Where changes to the industry were necessary Brunat's proposals were modest. Although the government, at least intermittently, appears also to have viewed Tomioka as a business venture, officials frequently ignored advice that would, at the very least, have reduced initial capital expenditures and increased productivity.

From a technological perspective Tomioka was irrational: its technology was too expensive and overly complicated. From an economic perspective, in addition to the high initial capital investments required, the proposed filature was beyond the existing local market mechanisms in the 1870s. Simply put, the facility required too much in the way of raw materials. Local cocoon suppliers were unable to provide Tomioka with enough cocoons for it to run at full capacity, thereby adding to its inefficiency. Partly because the filature was located in a very remote area, still largely inaccessible to this day, the extra costs involved with importing raw materials from other parts of Japan created further problems: Tomioka's managers would not have had the financial resources with which to hire the additional workers necessary to bring the plant up to full capacity (assuming it had the raw materials).[62] It is clear that external conditions such as availability of raw materials, geographic location, and labor did not figure heavily in the government's plan.[63] As would be expected, with the exception of a few periods of severe economic entrenchment, the facility rarely turned a profit during its first decade of operation.

AN ALTERNATIVE THEORY ON *CHOICE OF TECHNIQUE*

If Tomioka was neither technologically rational nor economically sound, it had to have been established on some other basis. McCallion has correctly argued that part of Tomioka's problem was that the government never had a clear vision, let alone plans, for the facility. Tomioka's proposed mission was for it to serve as a model of industrialization that private reeling concerns could follow to improve the quality of Japanese silk. At other times, Tomioka also was supposed to operate as a profitable business venture. Whereas these goals are not necessarily mutually exclusive, they were problematic for the Meiji government. From what evidence remains, however, this may not have always been the case.

It appears that the government's original ideas for a publicly operated, model reeling facility were more in accord with what Brunat had suggested, but that somewhere along the line, ideas about Tomioka's purpose changed. Odaka Atsutada stated that he, Sugiura, and Brunat traveled to Maebashi some time around July 1870 to visit Hayami Kenzō and ask about Mueller's plans for the facility. He noted that at the time, Mueller's plans for Maebashi were more or less the same as what the government had in mind for Tomioka. At least according to Odaka's recollections, it appears that the Meiji government originally did not anticipate their first entrepreneurial venture into the silk industry to be anything more than a relatively humble facility.[64]

Perhaps Brunat's suggestions that Japan's silk industry would benefit from the importation of European steam reeling techniques coupled with Shibusawa's and Sugiura's memories of Paris and the prestige attached to winning an award at the international exhibition were enough to change what appears to have originally been a relatively modest venture into an ideal to which few could aspire and none could imitate. While we may never know for certain whose suggestions led the Japanese government literally to import a French filature, Shibusawa Eiichi is the likely candidate. In an address evaluating the progress of Japan's silk reeling industry, Odaka Atsutada stated that the *Ōkurashō* used Shibusawa's proposal for Tomioka.[65]

Odaka's praise of Shibusawa and Shibusawa's comments regarding traditional Japanese silk reeling methods are also revealing. Shibusawa was a strong critic of Japanese silk reeling techniques. He characterized them as imperfect and recommended the adoption of European methods. In an address given in the mid- to late-1880s Odaka mentions Shibusawa's earlier statement that "the Japanese way of reeling silk was deficient (compared with that of the West)."[66] Odaka went on to praise the progress made by Japan's silk reeling industry in recent decades. He noted that the foundations of Japan's modern industry relied on the efforts of Shibusawa Eiichi.[67]

There would seem to have been a third function for Tomioka that helps explain why the government deviated from its original plans and, more importantly, helps explain why government officials attempted to transfer a technology without considering factors such as those mentioned above. It also provides some insight into why the government rejected the available alternative Western reeling technologies. Tomioka was an ideal. It was designed to be the physical manifestation of the (new) central government's authority over the *han*, and it was to serve as an exemplar of Western "civilization" in Japan. In the quest for "progress" and "civilization," whose model better to follow than France (and Britain), the key political and economic players, who were also the archetypes of *bunmei kaika* materiality.

At the time Tomioka was being conceived, the Meiji government's political position was anything but assured. When Itō was first approached in 1869 to establish a filature using Western technology the daimyō still had not, even symbolically, "returned" their domains to the emperor. In 1870, when Hayami established the Maebashi filature, it was under the authority of Maebashi-*han*, not the central government. In fact, the *han* were abolished six months *after* construction on Tomioka began. Umegaki argues that the first few years following the restoration of imperial rule were marked by a series of political moves that could best be described as a "simultaneous dispersion and consolidation of political power."[68] In an effort to stabilize the country, members of the new government on the one hand sought to consolidate its political power vis-à-vis the *bakuhan* system, and on the other hand sought to invest the domains with constitutional equality in order to eliminate inter-*han* rivalries.

The new government's change in plans for Tomioka reflected its need politically to situate itself above the *han*, and more importantly, the now defunct *bakufu*. After all, Western-style factories had been part of Satsuma-*han*'s industrial landscape as part of its efforts to industrialize the cotton industry in the decade before the Restoration; and it was under Tokugawa authority in 1865 that French engineers built the Yokosuka ironworks, Japan's first official attempt to import Western industrial technology under the supervision of foreign advisers. Toward that end, any industrial venture sponsored by the new central government would also serve the purpose of legitimizing its political authority.

The Meiji government also needed to demonstrate that it was the legitimate heir to Tokugawa political authority to the Western powers whose highly visible commercial and quasi-diplomatic activities in the treaty ports kept the bureaucracy on edge. The problem was not simply a matter of assuring the West that the Imperial government was in control. It was also trying to demonstrate to the West that Japan was indeed "civilized." On the popular level, "civilization" appeared in Japan through the adoption of Western clothes, hairstyles, and other insignificant, although easily observed and criticized, gestures.[69]

For the government the task was greater. By importing a full-blown modern French factory, the new government believed it was importing Western civilization to Japan. The *Minbushō* was also not alone in its belief that adopting material representations of the West would be representative of "civilization" and "progress" in the country. Writing in the summer of 1872, Inoue Kaoru, then head of the *Ōkurashō*, expressed his frustration to Kidō Takayoshi about the economics of this type of thinking. He complained about excessive capital expenditures: "regardless of the manner by which we choose to raise [our] level of civilization, there will not be any way to make use of it."[70] The following January he again made clear his frustration to Kidō: "Although the Finance Department alone has tried to limit government spending, the other departments, with the idea of gaining parity with the West, are insisting on [their] positive policies."[71]

At some point between July and November 1870 something changed in the government's thinking about what type of facility Tomioka was to be. The alternative technologies, such as those found at Maebashi, were capable of producing high quality silk. However, their use of traditional materials, that is, wood, and unsophisticated appearance made them unacceptable from an ideological perspective. During his visit to Maebashi filature in July 1870, Odaka betrayed this very point. He expressed dismay at the "rickety" wooden machines of which Mueller and Hayami appeared to be so proud.[72] If one's mission was to raise the level of "civilization" in Japan vis-à-vis the West, then only the most modern facilities made from the most modern materials were acceptable.

Economic and technological considerations do not seem to have been a concern. After all, Odaka and Brunat had no doubts about the abilities of Maebashi's alternative technologies to produce high quality silk. Nor were they concerned as to the ability of homemade reeling apparatus to be successful. As part of their evaluation of Maebashi's technology, Odaka and Brunat had four experienced workers reel silk for 30 days on machines described by Odaka as "Japanese-style modified to the least extent possible to make them European-style." At the end of the trial period the machines and silk were said to be of high quality and equal to those of Maebashi.[73] For

Odaka and Brunat, economics were a consideration. They were both concerned about the expenses involved with establishing the model filature, although their caution and worries were disregarded higher up the line.

Shibusawa's retrospective appraisals of Tomioka are indicative of beliefs in the superiority of Tomioka's technology and the facility's greatest value. They illustrate what one may argue was *the* determining factor in Tomioka's design: ideology. Writing in the mid-1930s, for example, Shibusawa stated:

> Although there may be some parallel example [of a reeling facility], nothing rivaled Tomioka in terms of being on such a grand scale and so perfect and complete. Its reputation rose to the world's attention and it was admired as a model facility from thriving cities to the countryside.[74]

Shibusawa's comments are similar to those of government officials who were opposed to selling the facility during the Matsukata Deflation. In 1881, a bureaucrat in the *Nōshōmushō* argued that Tomioka must be maintained because it was a source of international prestige for the Meiji government. As early as 1873 this was also the case. Sano Tsunetami's report on that year's International Exhibition in Vienna clearly describes the honor and prestige that the medals won by the model filatures had brought to Japan. He states that the "kankōryō filatures' silk is of the best quality and to look at its thread is to *see progress*."[75] Time and again, government officials ignored, or at least rationalized, Tomioka's financial and operational difficulties because of its prestige value. With this prestige came the belief that Japan's industrial progress as embodied in Tomioka was helping the country become "civilized."

Conclusion

When the Tomioka Silk Filature opened its doors in 1872, it was the product of neither careful technological nor economic evaluation. More accurately, alternative technologies were evaluated and the cost was considered, however, the advice the government received was largely ignored. At its inception Tomioka was to be a modest venture, probably nothing more elaborate than the small mill Hayami Kenzō had built at Maebashi. This is also largely reflected in Brunat's original proposal to the government. Brunat and Odaka evaluated Maebashi's (alternative) technologies and were favorably impressed with the quality of the silk produced. They even conducted their own experiments on locally manufactured machines to see if they could duplicate or surpass what was being done at the former *han*-based enterprise.

Brunat's proposal, while arguably written from the dual perspectives of establishing a profitable business and improving the quality of Japan's raw silk, considered any number of factors that reflected his training in Lyons and the state of the industry in general. He recommended modifying local technologies with recent European innovations only to the extent that it would not disrupt the abilities of local reelers to function in their craft. At the same time, if his recommendations were followed, it would have brought Japan's silk industry to the level of many European reeling facilities. This was also in accord with a directive issued by the *Minbushō* in February 1870, which called for reelers and merchants to set aside their greed and build European-style reeling machines in an effort to improve the quality of Japan's raw silk.[76]

In the months that followed the decision to build the facility, the government's thinking changed. Not only would Tomioka serve as a model of mechanization for local reelers, but it would also, more importantly, be an exemplar of "civilization" in Japan. Along with caution, Brunat's prospectus was abandoned as he was ordered to import and build a full-blown French filature in the backwoods of Gumma prefecture. This decision was based on the recommendations of Shibusawa Eiichi, who had been favorably impressed with France, Lyon silk, and the prestige it brought that nation at the Fifth International Exhibition in Paris in the spring of 1867.

The government's lack of caution, its alternative basis for choice of technique, was largely a function of the times. In an effort to assert its position over the *han*, the new government needed a model facility that placed it well above these semiautonomous territories, which were still vying for political influence. The new Imperial government also needed to demonstrate to the international community that it was the legitimate heir to Tokugawa political authority. Toward this end, it used the material trappings of the West to its best advantage. Tomioka Silk Filature would represent the march of Western progress and "civilization" in Japan. For the same reasons that a section of Tokyo, the Ginza, was completely rebuilt in brick after an 1872 fire, as evidence of "civilization" and, to quote Saigo Takamori, "for the honor of Japan," Tomioka was designed on the basis of symbolism.[77]

The Meiji government's first industrial venture was not part of any grand plan. It was based on a process in which the components of mechanization were gradually incorporated and assimilated to accommodate the government's varied demands. The primacy of technological development and improving the quality of raw silk gave way to economics; but later economics succumbed to ideology. In the end, the new government had its filature, however ill defined and unstated its purpose. Because Tomioka and the government's efforts are frequently evaluated in terms of financial success, the facility is often judged a failure. This myopic view of Japanese industrialization, however, fails to take ideology into account, and was I believe the fundamental factor in the choice of technique. As the Tomioka model demonstrates, beliefs in the ability of transferred technological artifacts to impart cultural and social values are critical to their selection.

NOTES

1. Stephen McCallion, "Silk Reeling in Japan: The Limits to Change," unpublished Ph.D. diss. (The Ohio State University, Columbus, 1983), pp. 84–92.
2. Ōkurashō, *Kōgyō iken* (Tokyo: Ōkurashō, 1884); reprinted in *Meiji zenki zaisei keizai shiryō shūsei: Ōkurashōhen*, Vol. 18 (Tokyo: Meiji Bunkan Shiryō Enkōkai, 1964), pp. 35–40. Also see, Nakamura Naofumi, "Kōhatsu koku kōgyōka to

32. See Silvana de Maio, “Italy, 9 May–3 June 1873” in *The Iwakura Mission in America and Europe: A New Assessment* ed. Ian, Nish (Surrey: Japan Library (Curzon Press Ltd.), 1998), pp. 149–161, on pp. 149–151.
33. Ibid., p. 155.
34. Ōtsuka, *Sanshi,* pp. 320–322.
35. *SYZ*, pp. 373–381. The evaluations of Italy are based only on Shibusawa’s recollections, Sugiura had returned to Paris and ultimately Japan the month before.
36. McCallion, “Silk Reeling in Japan,” p. 82.
37. Ōtsuka, *Sanshi*, pp. 256–257.
38. Ibid., p. 257.
39. Odaka Atsutada, “Seishi no hōkoku,” *Ryūmon zasshi,* May 15, 1893, *60*: 1–15, on p. 5; excerpts of this article are also reprinted in *SEDS*, Vol. 2, p. 524. Odaka stated that there were 36 frames, Ōtsuka’s description for Maebashi counts 32 reeling frames.
40. McCallion, “Silk Reeling in Japan,” p. 134.
41. The first group of six women were part of the workforce from a newly established filature in Tochigi-ken that also relied on “Western-style” machinery. In June 1871 visitors from Kumamoto-ken came to Maebashi to learn mechanized reeling techniques. Similarly, a group from Shinshū also came to study at Maebashi. This group was part of a request to set up five additional facilities in Shinshū. Ōtsuka, *Sanshi*, pp. 260–261.
42. McCallion, “Silk Reeling in Japan,” pp. 263–268.
43. Kannōkyoku, Shomukyoku, *Kyōshinkai hōkoku, kenshi no bu* (Tokyo: Yūrindō, 1880), p. 39.
44. Ibid., pp. 24–89 passim.
45. McCallion, “Silk Reeling in Japan,” p. 267.
46. See Ōkurashō, *Kōbushō enkaku hōkoku*, Tokyo, 1889, pp. 684–688. The Akasaka filature was also known as the Tokyo Kankōryō Seishijō (filature).
47. Ōtsuka, *Sanshi*, p. 317.
48. Sano Tsunetami. *Ōkoku hakurankai hōkokusho: sangyōbu*, Tokyo, 1875, Vol. 1, section 6, p. 7.
49. Ibid., Vol. 2, section 1, p. 7 (emphasis mine).
50. Motoyama Yukihiko, trans., George Wilson, “Meirokusha Thinkers and Early Meiji Enlightenment Thought” in *Proliferating Talent: Essays on Politics, Thought, and Education in the Meiji Era* (Honolulu: University of Hawaii Press, 1997), pp. 238–273, on p. 239.
51. Gregory K. Clancey, “Foreign Knowledge or Art Nation, Earthquake Nation: Architecture, Seismology, Carpentry, the West, and Japan, 1876–1923,” unpublished Ph.D. diss. (Cambridge, Mass.: Massachusetts Institute of Technology, 1998), p. 13. Brick chimneys were also considered representative of the march of industrial progress, p. 65.
52. This fact was not lost on Shibusawa who marveled at the steam engine at the British exhibit in 1867.
53. Muramatsu Teijirō, “Basuchan-gunkan to kinu no ito,” *Oyatoi gaikokujin,* Vol. 15, *kenchiku-domoku* (Tokyo: Kashimada Shuppankai, 1978), p. 129.
54. Odaka Atsutada, *Rankō ō*, Tokyo, 1909, pp. 204–206
55. Nagai Yasuoki, *Seishika hikkei,* 3 vols., Tokyo, 1884. The first release of the manuals was in 1878. For details regarding filature construction, see *2*:12 ff.; for reeling techniques, see 3: 1–31.
56. Ibid., vol. 2, p. 3.

57. Ibid., p. 4.
58. Ibid., p. 12.
59. See Federico, *An Economic History of the Silk Industry*, pp. 106–107; and Yukihiko Kiyokawa, "Transplantation of the European Factory System and Adaptations in Japan: The Experience of the Tomioka Model Filature" *Hitotsubashi Journal of Economics* 1987, *28*: 27–39, on p. 29.
60. Nagai, *Seishika hikkei*, Vol. 3, p. 14.
61. Penelope Francks, *Technology and Agricultural Development in Pre-War Japan*, (New Haven: Yale University Press, 1984), p. 11.
62. *TSS* document no. 233, *1*: 539.
63. Odaka blamed Brunat for choosing the site on which to build the government's model facility. Brunat's reasons were stated to be that the location would not cause hardship to local residents, although some had to be relocated (they were well compensated for their land). Odaka's complaint was not with the remote location, but with the water supply, a key ingredient in silk reeling. Odaka, *Rankō ō*, pp. 199–202.
64. Odaka, *Ryūmon zasshi*, p. 4.
65. Ibid., p. 9. "Shibusawa's proposal" would be the terms of construction elaborated in Brunat's contract which included importing the reeling equipment, steam boilers, and steam engines from France, the government's determination of Tomioka's architecture, and the use of coal for fuel, as well as guidelines for hiring foreign and local workers, their salaries, and terms of employment. See *TSS* document 4, 1: 150–152.
66. Ibid., p. 3.
67. Ibid., p. 8.
68. Umegaki, *After the Restoration*, p. 15.
69. Vice Consul Martin Dohmen to F. O. Adams, *Commercial Reports from Her Majesty's Consuls in Japan, 1871* (London: Harrison and Sons, 1872), p. 56. The Japanese adoption of Western clothing was a constant source of comment and amusement for foreigners, especially the British, in Yokohama.
70. Inoue to Kidō, Summer 1872, in *Segai Inoue-kō den*, Vol. 1, pp. 520–521.
71. Inoue to Kidō, January 22, 1873, in *Segai Inoue-kō den*, Vol. 1, p. 523.
72. Odaka, *Ryūmon zasshi*, p. 6. "Rickety" is Odaka's choice of words.
73. Ibid. By this time Maebashi silk was considered to be high quality and demanded a high price in international silk markets.
74. Shibusawa Eiichi, "Seiin sensei denhakkō," Vol. 7, section 5, pp. 54–61 (1934–1948), also reproduced in *SEDS*, Vol. 2, p. 522.
75. Sano, Vol. 1, section 5, p. 3 and section 6, pp. 7–8. In this instance Sano lumps together silk produced by Tomioka and the Tokyo Kankōryō Filature. At other times he and the Austrian officials at the exhibition distinguish between the Tomioka filature and the Tokyo Kankōryō (Akasaka filature) in name only (emphasis mine).
76. The text of the directive is reproduced in Ōtsuka, pp. 249–250. Merchants and reelers were rightly accused of corner-cutting and selling inferior quality silk to take advantage of the newly booming silk export market.
77. Clancey, "Foreign Knowledge," p. 9.

7

A Miracle of Industry: The Struggle to Produce Sheet Glass in Modernizing Japan

Martha Chaiklin

The history of the Meiji period often seems to have been written with slogans such as "civilization and enlightenment," "prosperous country, strong army," or "good wife, wise mother," as if the vast and complex political and social changes that occurred between 1868 and 1912 could be tamed by imprisonment in precise phrases.[1] Perhaps supreme in the hierarchy of slogans is "rapid modernization." Few would question that Japan was able to industrialize, modernize, Westernize, and metastasize in an almost miraculously short period of time; yet, the very truth of this statement somehow demeans the struggle. Nowhere is this more apparent than in the sheet (or flat) glass industry, where the road to success was paved with huge capital losses by both the Meiji government and private entrepreneurs.

The reason so much effort and expense were poured into learning to produce sheet glass can be surmised in one word: windows. As Susan B. Hanley has pointed out in *Everyday Things in Premodern Japan*,[2] in many ways, the quality of life in Japan before the arrival of Commodore Matthew Perry in 1853 was as good as or better than that in the Western world. Traditional Japanese architecture had existed for centuries without relying on glass windows. As a result, some came to see "the beauty of a Japanese room depends on a variation of shadows, heavy shadows against light shadows," rather than brilliantly lit through sun-filled windows.[3] Fourteenth-century monk Kenkō suggested a basis for this aesthetic when he wrote, "A house should be built with summer in mind. In winter it is possible to live anywhere, but a badly made house is unbearable when it gets hot."[4] Some scholars have theorized that the post and beam architecture of Japanese houses led the floor rather than the wall to become the focus of attention.[5] Another theory was proposed by art collector and critic James Jackson Jarves:

> Instead of costly framed landscapes hung on their walls, the nobles make their rooms scrupulously clean, airy and spacious, with movable divisions or screens,

> which can be so arranged as to leave open, as if inclosed [*sic*] in frames, attractive vistas of out-door scenery.[6]

The most popular theory of all was that expressed by Christopher Dresser who suggested it was the "great danger in a land subject to earthquake shocks."[7] There is no doubt, however, that in general walls were designed to move rather than have holes cut in them, so little use was made of the window as an architectural element.

Initially at least, the change from traditional to Western architecture was almost entirely political. The Western houses and shops in treaty ports were built for foreigners who owed their presence to gunboat diplomacy. More significantly, the Meiji government, aware that commercial treaties concluded in the 1850s were not to Japan's benefit, made treaty revision the main foreign policy goal through most of the second half of the nineteenth century. As the unfair treaties stemmed from Western attitudes of superiority, the Japanese government made many decisions to appear equal to narrow-minded Victorians, and thus be taken seriously. Real change was not always the desired result. For example, a series of appearance-related decrees was passed in the 1870s that mandated adoption of Western hair styles, Western clothing for government officials, and Western-based student uniforms. Nevertheless, most people continued to wear traditional dress.

Some nineteenth-century Westerners such as Lafcadio Hearn appreciated Japanese architecture, but the opinion of Captain L. L. Janes, was more common. He wrote:

> The impractical idealist, the *ennuye* tourist, and the briefed renegade from the higher civilization of the West, all in turn dilate on the convenient and beautiful "simplicity" of the Japanese homes and home-life. . . . Human progress, which is not only the destiny but the duty of the human race, is in its healthiest stages when in sympathy with the fundamental principles of evolution; when it is occupied in disintegrating and eliminating this maudling affectation of simplicity, and in reveling the higher beauty which inheres in perfectly adapted complexity.[8]

It was this kind of perspective that resulted in government buildings that were wholly Western rather than a hybrid style that allowed, say, the wearing of shoes in an otherwise traditional structure.

It was acceptable to construct Westernized homes and regional public buildings from wood, but the new government sought something more substantial for national edifices. While brick and stone were both options, red brick was favored in Meiji Japan as the less expensive and minimally more earthquake-friendly option. Unlike sheet glass, widespread knowledge of ceramics meant that the Japanese were able to produce bricks fairly quickly. The brick industry developed with almost no government support. Similarly, Portland cement, the most up-to-date mortar in the late nineteenth century, was in production by the second half of the 1870s. Sheet glass, however, was the one building material essential for producing Western buildings that was not available in Meiji Japan. While glass had been produced in Japan at least as early as the eighth century, the many important technological advances such as glass blowing and annealing (the process of slow cooling to increase durability) that had been made during the early modern period did not include any significant production of window glass. The basic methods for producing flat glass were supposedly introduced by the Dutch in the mid-eighteenth century but only small crude pieces were produced in very limited quantities.[9] These rough sheets were used primarily for mirrors but occasionally a small pane was inserted into a shoji screen, which were called "snow-viewing windows."

Scholars and the upper echelons of society were in fact fairly familiar with higher quality sheet glass. Significant quantities were imported throughout the early modern period, primarily for optical lenses. Some other interesting uses are recorded, however. Tokugawa Ieyasu had a glazed room constructed. Similarly, in 1688, Date Tsunamune had glass panes inserted into the doors of his Edo mansion.[10] One *nouveau riche* merchant of the seventeenth century, Itō Kozaemon, even placed goldfish tanks in his ceiling. Still, these examples were the exception rather than the rule; glazed windows of any sort were only for the very rich.

Imported glass did not stimulate domestic production of sheet glass because the appearance of a piece of glass revealed nothing about how it was produced. Forming a flat piece of glass by hand was so difficult, only the most skilled glass artisans could accomplish it successfully. Before mechanization, there were two kinds of flat glass; blown and cast. Blown flat glass was produced by two techniques known as the crown method and the cylinder method. Both required a very strong metal (the technical term for raw molten glass). The crown method, which is older, formed flat sheets from a large bubble flattened through the use of centrifugal force. This process required great strength; furthermore, the size of the pieces of glass that could be cut from the compressed bubble was fairly limited. However, the bull's eye that formed in the center where the glass blob was attached to the pontil was also cut into decorative panes. To produce glass with the cylinder method, a large tube was blown, then snipped and pulled flat. Although larger pieces of glass could be produced than with the crown method, the cylinder method also required great strength to handle the unwieldy tube and skill to avoid imperfections in the glass that caused it to shatter. This technique did not come into wide usage until the nineteenth century. Neither the crown nor the cylinder method produced glass that was really flat; both left a slight curve.[11] Plate glass is cast and then polished. The metal therefore must be very hard and without imperfections. All of these processes were so difficult that the development of mass-produced sheet glass was considered a crucial goal for nineteenth-century glass producers throughout the world.

After all, windows are an almost universal requirement for a permanent structure. Blocks of ice are even inserted into igloos for this purpose. Although alabaster, mica, and shells were all used for windows in the West, the use of glass dates back to the ancient Romans, who used glass windows in their baths, albeit only small, thick pieces. Expense and taxation meant, however, that glass windows did not become common until the eighteenth century. Glass was, and still is, the most desirable material for windows because it allows the most light yet maintains a high degree of insulation. While traditional Japanese architecture relied on screens to diffuse light, inserting paper screens into Westernized structures would have defeated the purpose of creating a building to comply with the Western aesthetic. Attempts to insert glass into a traditional Japanese aesthetic provided equally unsatisfactory. In 1933, novelist Tanizaki Junichirō wrote of the difficulties he encountered when attempting to combine shoji and glass in his new home:

> A few years ago I spent a great deal more money than I could afford to build a house. I fussed over every last fitting and fixture, and in every case encountered difficulty. There was the shoji: for aesthetic reasons I did not want to use glass, and yet paper alone would have posed problems of illumination and security. Much against my will, I decided to cover the inside with paper and the outside with glass. This required a double frame, thus raising the cost. Yet having gone to all this trouble, the effect was far from pleasing. The outside remained no more than a glass door; while within, the mellow softness of the paper was destroyed by the glass that lay behind it. At that point I was sorry I had not settled for glass to begin with.[12]

Few Westernized buildings were built before the Meiji Restoration (1867–1868). Homes for Western merchants were built in the foreign concessions within a few years of the ports opening for trade in 1858, and hotels and shops soon followed, but this was not representative of most of the country. The shogunate only initiated a few Western-style construction projects; iron foundries in Nagasaki (1857) and Yokosuka (1860) and the Tsukiji Hotel (1867), which was intended for foreign visitors. The hotel was not actually completed until the following year, after the Meiji Restoration. While the window glass for all of these projects had to be imported, the number of panes required was so low the expense was still acceptable.

It was, therefore, a Meiji initiative to produce Western architecture and windows, but it was not until the 1870s, after the civil unrest and organizational concerns had been resolved that the government could focus on architecture. Early government projects included the Osaka Mint (1869), the Yokohama and Shinagawa railway stations (1872) and the Yokohama Customs House (1873). The reconstruction of Ginza was, however, by far the largest Westernized government urban planning project in Meiji Japan. After a particularly devastating fire in 1872, the government decided to build a fireproof section of brick buildings to better complement the new stone-faced railway station and would showcase modern Japan. Designed with the advice of English architect Thomas Waters, the new cityscape had a tree-lined arcade surrounded by almost 1,500 Western-style brick buildings with glass windows.

It would appear that even with the limited experience that the Meiji government had with Westernized buildings before the Ginza project, there was an awareness that sheet glass would be an issue. In 1872, a delegation of top government officials, known as the Iwakura mission (because Prince Iwakura Tomomi headed it), traveled to the United States and Europe. Although the primary goal of the mission had been the renegotiation of treaties, when it became apparent that this was not possible, the mission set out to study the infrastructure and industry of the West as a model to bring prosperity to Japan. Mori Arinori, who designed the itinerary for the North American leg of the journey, did not schedule any mission visits to glass factories,[13] but in Europe the group visited several glassworks in England, Belgium and Italy. The Japanese party paid close attention to glass production methods, materials, and plant layouts, which were carefully described in the official published account of the journey. At St. Helens in Manchester they observed the production of plate-glass at one of the largest production centers of glass in all of Europe, and at Chance Brothers Glass and Chemical works in Birmingham, they saw cylinder glass being produced to make both sheet glass and signals for lighthouses. Here it was noted by Richard Henry Brunton, a former Japanese government employee, that "Ito [Hirobumi] received from him [either James or Lucas Chance] much valuable information regarding the manufacture of glass."[14] The group chose to observe sheet glass production over soda and sulfuric acid production, also produced at the same location. They also visited, Osler's, another glass factory in Birmingham. In Belgium, the party visited an "extraordinary" glass factory.[15] In Venice they saw beads and plate glass being produced, a visit that also included the Museo d'Arte Vetraria, or Glass Museum.[16] It is difficult to determine how involved the mission members were in choosing their itinerary but it seems apparent that the repeated visits and detailed, publicized commentary indicate a definite interest in glass production. In fact, the mission's official account decisively concluded that:

> To depend entirely on imports [of glass] is not in the interests of the economy and causes substantial inconvenience, moreover, so that sooner or later we must develop glass manufacturing ourselves.[17]

Government interest in sheet glass as an industry that would bring prosperity to the nation was tied directly to increased government construction. While Japanese philosophers expounded the philosophy of economic self-reliance as early as the seventeenth century, increased Western-style construction caused the reasoning behind this concept, that foreign trade would drain the country of its precious metals, to become a reality. Provisions in the commercial treaties such as the inability to control tariffs made foreign trade an economic drain on Japan. Duties were set at 20 percent on most items, with no special exception for glass. Since initially, the need for sheet glass was in the foreign concessions, and the Japanese were denied the right to set tariffs under the unequal commercial treaties, it was foreign demand that led to a reduced duty of only five percent as early as 1864.[18]

Samples of imported sheet glass arrived as soon as the treaty ports opened for foreign trade, but as most of it was for the foreign concessions, there was little incentive for Japanese traders to seek to take over the glass import business. For the better part of the nineteenth-century foreign firms handled all imports of sheet glass. Supposedly, the first to import glass panes was an Englishman, but firms such as Vigan & Company of France and Grösser & Company of Germany are also known to have handled glass.[19] Japanese companies did not become involved in glass imports until the end of the nineteenth century. This was partially due to the fact that the sophisticated domestic commercial structure that had developed under the shogunate did not train Japanese merchants to deal with foreign trade in a free trade system. Since the end of the seventeenth-century foreign trade had been handled through an office in Nagasaki known as the *kaisho*. All trade goods were funneled through the *kaisho* and then auctioned off to a group of merchants licensed to deal in foreign trade. After decades of these monopolistic conditions, Japanese merchants lacked foreign contacts and were utterly unprepared to handle the sudden changes brought on when the commercial treaties opened the ports. As a result, the glass passed through foreign middlemen before being distributed by Japanese companies. Changes in government policy concerning direct export subsidies in the mid-1880s forced many into bankruptcy or to move into other fields.[20] Even as late as 1887, direct imports by Japanese merchants only accounted for less than 12 percent of total imports.[21] Japanese firms did not get involved in importing sheet glass until a few years later: Mitsui Bussan, which was founded in 1876, was one of the first.[22] It was the modern trading incarnation of the House of Mitsui, a mercantile family that had begun its climb to economic supremacy through sake brewing in the 1630s.[23] Even in the Edo period, the House of Mitsui had maintained prosperity by diversifying into pawnshops, money lending and dry goods, including the Nagasaki imported textile trade. The family had survived the economic upheaval of the Restoration by obtaining lucrative contracts with the Imperial government. The Mitsuis were able to cement these ties by loaning the new government large sums when they were most desperately needed. The family established Mitsui Bussan as a broker for foreign goods established to, "export overseas surplus products of the Imperial Land, to import from overseas products needed at home, and thereby to engage in intercourse with the many countries of the Universe."[24] It gained a head start by taking over Senshu Gaisha ("Profits First Company"), a trading company founded by former finance minister Inoue Kaoru when Inoue was recalled to the cabinet. Despite the ambitious start, however, Mitsui Bussan did not begin directly trading abroad until 1882.[25] It was another decade or so before they imported sheet glass.

Because no window glass was produced domestically, increased construction resulted in ever-rising sheet glass imports. For the first five years after the Restoration, imports increased rapidly from 10,144 sheets in 1868 to 42,586 in 1872. The following year imports accelerated to 101,337 sheets. In 1875, Tsuda Mamichi (1829–1903), noted in an essay on trade imbalance that:

> As the lower classes invariably wax enthusiastic over what their rulers admire, the people in our empire are coming on the whole gradually to admire foreign ways, to wear foreign caps and clothes, and to construct foreign houses. Almost all the homes have come to depend upon foreign imports for household appliances ranging from such items as glass, mirrors, pictures, chairs, and tables to cakes, wines, and other edibles. Such are the reasons why the excess of imports over exports has reached the large sum of Yen 8 million each year.[26]

While sheet glass imports fluctuated up and down, there was a steady cumulative increase. Tsuda would have been shocked to know that by the time of his death, sheet glass imports had quadrupled. In 1894, glassware imports of all types were twenty-first in import value.[27] By 1901, demand had increased to such an extent that sheet glass alone was twenty-fifth in import value, exceeding a million yen.[28] In volume, the million-sheet mark was passed in 1900. Constantly escalating demands led Diet member Nemoto Masaru to produce an official government recommendation in 1902 that encouraged government support of window glass production. The statement read:

> Cotton is imported in massive amounts from America and India, but this is turned into a finished product and exported abroad creating a profit balance, but this is not at all the case for window glass; all imports are consumed domestically. The window glass imports in this past year were more than 1,250,000 yen . . . When compared to the imports of seven years ago of 165,000 yen they have increased eightfold. It is likely that the amount of window glass use will be a factor in determining national progress. Therefore, if we are unable to produce it in our country, in seven years hence, based on the seven years previous, imports will exceed 8,820,000 yen and could easily surpass 10 million yen. For this reason, this industry is not only a present need for our country at the present time, but there are few raw materials available domestically. The government should give significant protection and plan for the advancement of the domestic glass industry.[29]

At the end of 1902, the government decided to provide 501,000 yen assistance over four years for companies having one million yen capital or more. While a million yen was earmarked for this purpose, the funds were diverted by the need to fund the Russo-Japanese War. As a result, more than two million sheets were imported in 1906, and subsequent imports never dropped below two million sheets for the rest of the Meiji period.[30] Glass imports to Japan were primarily from Belgium, then the global leader in sheet glass production.

Even though the initial jumps in imports were created by government demand, the first attempts to produce sheet glass were in the private sector. Traditional glass producers were devastated by the influx of imports because they could not compete in quality or price. Most Japanese historians, however, have associated this downfall with a total break from past traditions.

However, in 1860, glassmaking was not fully mechanized in the West and still relied on skilled craftsmen. While the structural organization of a workshop was vastly different from a factory setting, the industrialization of the glass industry had to proceed with traditionally trained craftsmen.

Those such as Tokugawa Nariakira of the Mito domain recognized the need for sheet glass well before the Restoration.[31] Nariakira may have been distracted by the movement to restore the emperor so little came of his efforts to organize production. Kōgyōsha Glassworks, founded in 1873 by Niwa Masatsune, formerly steward to premier Sanjō Sanetomi, and Murai Mishinosuke, left a much greater mark on history than the efforts in Mito. Murai, who acted as the technical chief, was really the key figure in founding the enterprise. He thoroughly researched the Western glass industry and was able to convince Niwa to join him as factory head. Through Niwa's influence, Sanjō also became interested in the project.[32] Later known as Shinagawa Glass for its location in Tokyo, the Kōgyōsha factory was staffed with four glassworkers from Osaka and Tokyo (both traditional glassmaking centers) who had no previous experience with sheet glass. Part of the 200,000-yen start-up capital went to importing from England clay for the crucibles and firebrick for the furnaces. Erasmus Gower, a British mining engineer, was employed to locate domestic sources for quartz and silica.[33] The factory building itself was red brick with, appropriately, glass windows. The investors even brought a craftsman from England, Thomas Walton, to instruct the workers how to make sheet glass. Walton achieved very little during his tenure in Japan.

Traditional Japanese glassmakers used a lead-rich metal that was too soft and heavy to produce cylinder or plate glass. Under Walton, Kōgyōsha managed to produce flint glass metal, made with ground flint, which is hard and very clear.[34] This seems to have given rise to the idea that the flint glass was used for windows, but the high cost and weight make it unlikely the flint glass was ever used for window glass. It was probably developed to make ship signal glass because flint glass has superior refracting properties. Soda glass, the primary material for window glass throughout the world, had been produced in small quantities in Japan since about the mid-eighteenth century when the Dutch introduced the process. It is therefore unlikely that difficulty in producing the metal or the unfamiliarity of the craftsman in working with it were the problem. Rather, the difficulty lay in technique. Consequently, only a few pieces of flat glass were ever produced under Walton.[35] Kōgyōsha failed in just three years because huge investments had not resulted in the commercial production of sheet glass, the only product with a high enough profit margin to have made the factory viable. Niwa and Murai sold the factory to the Ministry of Technology (Kobuchō) in 1876.

Renamed Shinagawa Glass[36] under the government, there were still high hopes for success. In 1879, architect Sone Tatsuzō wrote about Shinagawa Glass in his graduation thesis from the first graduating class in architecture at the Imperial University. The topic, assigned by British architect Josiah Condor, was "The Future of Domestic Architecture in Japan." After acknowledging that Japan was still relying on imports from Europe, he wrote, flat glass would "probably eventually be produced in the flint glass factory at Shinagawa" and that he was looking forward to the day when the glass made there would replace imports.[37] In order to generate income, after 1877 the factory began to pursue the production of other kinds of glass necessary for modernization along with sheet glass, and the flint glass introduced under Walton resulted in commercial production of red signal glass for ships. Ships glass had been produced in Satsuma since 1858, when red glass was first successfully manufactured in Japan, but the Shinagawa factory used more modern methods.[38] Thomas Walton remained employed under the new management but much of the design work was carried out under Fujiyama Tanehiro.

Fujiyama was a prime example of a craftsman who had successfully made the transition from traditional to modern glassmaking. He began his career in Saga, in the domain-sponsored glassworks, one of the many enterprises initiated by daimyo Nabeshima Kansō in order to improve finances. Very little is known about what kind of glass was produced, but Fujiyama must have been a fine craftsman because Meiji government was confident enough in his skills to send him to study technique in the Stolzenfels factory in Austria, where he also studied printing and pencil manufacturing processes. While not the goal the government was striving for, small pieces of colored, patterned glass were produced under Fujiyama. After this small success, Fujiyama went on to pencil production, becoming the first pencil manufacturer in Japan.[39]

Thomas Walton went back to England in 1879 and was replaced by another English craftsman, James Speed. Speed was reputed to have been a higher caliber craftsman than Walton.[40] In an attempt to make the factory more profitable, tableware was produced under his instruction, eventually of a quality sufficient to be exhibited at Second Industrial Exposition in 1881. Later, another Englishman, Emanuel Hauptmann was brought to teach cut glass and engraving techniques.[41] The government had not given up on the idea of producing sheet glass, however. More capital was poured into the enterprise, some of which was for new furnaces that were built of firebrick imported from England. The new furnaces were completed in 1881, but no sheet glass was produced. Out of funds, the Ministry of Technology petitioned the Ministry of Finance for 94,000 yen to train craftsmen and conduct experimental production. They were granted 20,000 yen but these funds were eaten up in a matter of months. Plagued by quality issues, frequent management changes, and an inability to produce acceptable sheet glass, the Shinagawa glass works had a deficit of 200,000 yen when manager Abe Hiroshi successfully petitioned the government sell the factory to private investors. In 1884, Shinagawa Glassworks was purchased by two, investors Inaba Masakuni and Nishimura Katsuzō, for 79,950 with installments to be paid over 55 years.[42]

Nishimura was avid to make the factory a success. When he learned of the new of the Siemens furnace, he first went to Germany to observe it, and then sent craftsman Nakajima Sen to study at a factory there. Invented by Karl

Wilhelm Siemens (1823–1883), this wide, shallow, coal gas-fueled furnace used preheated air and exhaust gases that passed through closed brick chambers, which prevented heat loss from the chimney to achieve exceptionally high temperatures.[43] The Siemens furnace was first put to industrial use in 1861 at Chance Brothers, the same one in Birmingham visited by the Iwakura mission. Before his return to Japan in 1888, Nakajima ordered some molding and pressing machinery and a Siemens furnace. He also brought back two German glass artisans. With the new machinery and craftsman, Nishimura was able to mass-produce beer bottles at 350 bottles per craftsman per day—about four times the output of other Japanese craftsmen, for a total of 8,000 to 9,000 bottles a day. He was thus able to make the factory turn a profit for the first time, finding a steady customer in the Japan Brewery Company of Yokohama. With this success, company capital quadrupled to 600,000 yen, which encouraged the two entrepreneurs to attempt to produce sheet glass. Two years were devoted to preparations but the timing was inopportune—a nationwide depression dried up the profits that were to be used to start up the new branch. Lacking a buyer, the Shinagawa Glass factory closed its doors in 1892.[44]

Despite massive investments from both the public and private sectors, the only time the Shinagawa Glass Factory was ever out of the red was for that brief period under Nishimura when beer bottles generated income. There were two causes for this; in a country where high quality porcelain was widely available and the primary drink was tea, glass tableware was not in high demand; but the more significant problem was the factory's inability to produce sheet glass. Shinagawa was not a success for its investors but it was not a total failure. Although sheet glass remained elusive, many new glass products were produced in modern industrial conditions. Perhaps even more importantly, many employees at the factory, given the opportunity to work with foreign craftsmen, went on to work in other enterprises, thus sowing the seeds for industrial development.

Itō Keishin, employed many Shinagawa veterans in his Osaka start-up, Nihon Garasu Gaisha (Japan Glass company). Little is known about Itō, but it appears he was attracted to the glass industry in 1875 because he hoped to shift the profits from imports to himself. Keishin built a furnace with the assistance of Osaka Mint engineer Shimada Bo.[45] Inexperience resulted in many failures, but Itō persisted with research and tests. Five glassblowers, Noda Shinzō, Shino Inosuke, Shimada Magoichi, Ikeda Sukeshichi and Yamada Eitarō, were all brought to Osaka from the Shinagawa factory to provide expertise. Itō's persistence paid off with the successful production of ships signals, retorts and acid-proof bottles. In 1881, he applied to the government for funds to expand but was turned down. Nevertheless, he incorporated that year with capital of 180,000 yen. He designated Hata Kyōsuke as president and himself as head engineer. The following year (1882) he brought an English craftsman named Elijah Skidmore from Shinagawa to make crucibles from local clays rather than the imported clays that had been used over the past few decades. Itō traveled all over Japan to find a suitable clay source, almost exhausting his resources before discovering some in Shigaraki. As a result, Nihon Garasu also went on to produce firebrick. Emboldened with success, Itō decided to pursue sheet glass. However, deemed responsible for past financial difficulties, Itō was forced to leave the company in 1888. He remained in the glass industry until his death in 1900, but was never able to make a company profitable.[46] Itō Keishin and Nihon Garasu are important for the same reasons that Shinagawa Glass was—they paved the way for the companies to come, providing training for craftsmen and introducing Western technology and techniques. Nihon Garasu was also important because it established industrialized glass production in Osaka. While successful production of sheet glass had not yet been attained in Osaka, other glass products were "made with economy and success" by the 1890s.[47] Osaka was later to become the center of the sheet glass industry in Japan.

Iwaki Tasujirō was a Shinagawa veteran who founded his own company, Iwaki Glass, in Tokyo in 1881. He had become adept at producing lamp chimneys and other blown glass but he too was tempted by the grail of sheet glass. To achieve this goal Iwaki went to America around 1896 to study American sheet and other glass production methods. He returned in 1900 and, with the addition of 65,000 yen from a group of investors, purchased the old Shinagawa Glass Factory facility in March 1900. He manufactured his own firebrick and his own furnace and about four months later, a sheet of glass about six foot square was produced.[48] However, the failed attempts that had preceded the success had cost more than 7,000 yen. Moreover, it cost 100 to 150 yen to produce each piece. This was at a time when skilled laborers such as carpenters earned about 160 yen a year, and a male contract farm laborer only about 32 yen annually.[49] By this time, the government was no longer assisting private enterprises beyond the loan of equipment. Iwaki's death in 1915 caused the sheet glass division to shut down with heavy losses because his experience and skill could not easily be replaced, even with the help of an infusion of 30,000 yen capital from Dai-ichi Kangyō Bank.[50]

While the Ministry of Technology had refused to support Iwaki, interest in the glass industry was present in another branch of government, the Ministry of Agriculture and Commerce. Takayama Kentarō, a doctor of engineering who headed a government research station, submitted a report to the government in 1905 recommending the establishment of a test facility for the production of sheet glass. In this report, he pointed out that private industry could not be relied upon to provide the funds for training craftsmen, a process that could take several years. He pointed out that if things were left as they were, entrepreneurs would avoid the industry and imports would only increase. It was his opinion that it was the government's duty to create this test facility to educate craftsmen so that they could be employed by private industry. With support from several key government officials, Takayama was given the go-ahead to start planning. He began by requesting assistance from Hirano Kōsuke who was studying in Germany. Hirano was charged with finding out about the newest furnaces and signing on a European craftsman to instruct the Japanese workers. Hirano and other government employees were also sent to Belgium, then the world's leading sheet glass producer, to study technique there. In addition to technical information, the detailed report Hirano wrote about his journey debated whether sheet glass production should be a government or private enterprise (see figure 7.1).[51]

Figure 7.1 Production by the cylinder method.

Source: Asahi Glass Co., Ltd.

Ultimately private enterprise triumphed when Shimada Glassworks, owned by Shimada Magoichi, was able to bring the first domestically produced sheet glass to market in 1903.[52] Shimada was a veteran of the glass industry. He had begun in a traditional workshop, apprenticed at Shinagawa under James Speed, and worked again under Itō Keishin, finally going independent in 1882. His commercial success with other glass products and his extensive background, made it difficult for Shimada to resist the call of sheet glass. In 1893, he purchased Itō Keishin's factory for 27,000 yen. Success came just under a decade later. Magoichi submitted the sheet glass he produced in 1902 to the Fifth Domestic Industrial Fair the following year and took a prize. He was finally able to bring it to market in 1904, although there were still many quality issues.

Eager to improve and expand his production capacity, Shimada Magoichi, approached Iwasaki Toshiya to suggest a joint venture (see figure 7.2).

Iwasaki, who was related to the Mitsubishi zaibatsu family, had studied chemistry in London specifically to enter the field of glass production. With 450,000 yen capital from Iwasaki and the factory, valued at 300,000, the Osaka Shimada Glass Company was formed in 1906, but it did not fare well. Decreased demand of the previously successfully glassware line accompanied

Figure 7.2 Iwasaki Toshiya.

Source: Asahi Glass Co., Ltd.

by disagreements between the young Iwasaki and the much older Shimada, which soon led to a split in 1908. Shimada went back to his roots, producing glass through most of the twentieth century.[53]

Iwasaki went on to found his own company, Asahi Glass in 1907, with a downsized capital investment of 200,000 yen. Iwasaki set up his new venture in Amagasaki, in Hyogo prefecture (see figure 7.3).

The new factory, designed by a Belgian and utilizing Belgian methods to blow glass, produced commercially viable sheet glass only two years after commencing production, but only with great difficulty. In 1909, Iwasaki brought over five Belgian craftsmen and obtained another one, a Mr. Helman, from the failure of another start-up, the Tōyō Glass Manufacturing Company.[54] Commercial production of sheet glass was at long last accomplished in 1910 but only enough to produce four percent of domestic glass needs, or about 60,000 boxes. All raw materials were local, except soda ash imported from England. Despite this modest success, Asahi was still struggling to produce sufficient quantities of adequate quality glass. In an attempt to remedy this, the Asahi Company introduced the Lubbers process. This was a semi-mechanized form of cylinder glass invented in America about 1896. Pressurized air was pushed through a blowpipe and the resulting bubble was then forced into guides to produce cylinders. The Lubbers process made production somewhat easier but still required laborious hand flattening. The German invasion of Belgium in 1914 was a boost to Asahi because Belgium had been a major

Figure 7.3 Amagasaki Factory.
Source: Asahi Glass, Co. Ltd.

glass exporter on the international market. Prices rose for sheet glass and the company was actually able to export window glass that year to England, Australia and all over South and Southeast Asia.

Asahi was the only Japanese commercial producer of sheet glass from until the founding of the America-Japan Sheet Glass Company. In 1917, Sugita Yosaburo concluded an agreement with Libbey-Owens Sheet Glass Company to obtain the Japanese patent rights for the Colburn method.[55] Previous generations of the Sugita family had been in the firewood business, supplying fuel for the copper refinery run by the Sumitomo family. This business relationship continued after the Restoration, with the Sugita family supplying fuel for other Sumitomo mining interests. Sugita was selected to negotiate the patent rights because he had extensive experience abroad and in foreign trade. He spent six years studying in the United States, graduating from the University of Chicago in 1906, followed by a study of commerce and industry in New York and London. In 1912, he entered the export division of the Shima Trading Company.

Sugita's relationship with Libbey Owens began before the Libbey Glass Company had even merged with the Owens Bottling Company. Sugita was first attracted to glass through a deal to supply beer bottles to Hong Kong. He was shocked at the primitive state of production at his supplier, Osaka Bottle Factory, and the bottle industry in Japan as a whole. Uesugi Senpachi told Sugita that the situation could be remedied by obtaining an Owens Bottle machine for mass production but the high patent costs made it impossible for any of the fledgling bottling companies in production in Japan to obtain one. Arriving in the United States in 1914, Sugita opened negotiations with John D. Biggers and William S. Walbridge of the Owens bottle Machine Company through a letter of introduction from his mentor, railroad man D.W. Cook. Sugita obtained the bottling patent for 300,000 yen—100,000 in cash, 100,000 in stock, and 100,000 as a loan. When Sugita got back to Japan, he found that Uesugi had died. Lacking a partner, Sugita convinced Shima to go into the glass bottle business.

While Sugita was in the United States, he observed experiments at the Toledo Glass Company involving the Colburn method. This was a mechanized rolling method that had been invented by Irving W. Colburn in 1908, and perfected by Toledo Glass Company and the Owens Bottle Company in 1916. The Colburn method was an advance over earlier methods because it rolled the glass in molten form rather than produce a cylinder through mechanization that still had to be rolled out by hand.[56] Sugita shrewdly awaited the perfection of the Colburn method. He then convinced his employer, Shima Trading, to allow him to negotiate the patent rights for the process when it went on line in 1916.[57]

While both sides were interested in the deal, negotiations faltered because Libbey-Owens wanted twice the 2 million yen allocated for patent rights in Sugita's budget. Sugita was persistent, however, and agreement was reached within two months, at the end of 1917. The Japanese were to pay Libbey-Owens $100,000 and a full one-third of the initially authorized capital stock.[58]

The agreement was null and void if a company was not launched by September 1918. Libbey was to supply all kilns, machines and other equipment as well as provide an instructor experienced with the new technology. The new Japanese company was forbidden from exporting any of the technology supplied by Libbey-Owens or the products produced with the Colburn process. Capital was garnered primarily from Sumitomo Bank and Asahi Glass. The joint enterprise, appropriately called the America-Japan Sheet Glass Company, Ltd. (Nichi-bei Ita Garasu Kabushiki-Gaisha), introduced the Colburn method to Japan. Today this company is known as Nippon Sheet Glass Company. Sugita was rewarded with 2,000 shares, or one-tenth of the stock. The next year, Sugita was made one of five directors of the new company.

In order to make the company a success, two craftsmen from Asahi, Yamada Naoichi, who was responsible for the production end, and Aoki Sakichi, who was responsible for the crucibles, were sent to the Charleston, West Virginia factory of Libbey-Owens. Tayama Kantarō and Suzuki Kaoru, recent graduates of the Tokyo Higher School of Engineering with no practical experience, were also sent. Construction of the factory began in 1919, and test production began by the end of the year. Finally, in October of 1920, production commenced. Unlike Japan's past forays into sheet glass production, success was immediate. The factory produced 1,525 boxes in its first year. The following year, engraved and ice glass were also produced at the factory.[59]

The America-Japan Glass Company was remarkable because it was a joint venture, and thus, in a sense, a departure from the Japanese doctrine of self-reliance. It was not the first attempt to produce flat glass through a foreign partnership. Just over a decade earlier, in 1906, the Tōyō Glass Manufacturing Company was formed to produce sheet glass, bottles and tableware with capital contributions from Belgian, British, and French companies. Japanese investors included Shibusawa Eiichi. Huge investments were made to bring over five Belgian craftsmen, as well as construction of a factory, quarters for the foreign workmen, a canal link, specialized furnaces for tableware and sheet glass and an annealing oven for sheet glass. In an apparent case of too many cooks, combined with obvious overspending, the company was liquidated three years later without having produced anything. Even this short-lived company was important because the factory set-up was wholly Western and it introduced the latest furnace technology.[60] Because Asahi absorbed the remnants, Tōyō was directly related to the success of Asahi and by extension, to the success of the America-Japan Glass Company. Glass was produced rapidly at the America-Japan Glass Company because the technology relied less on the skills of the individual worker, but also because the structure of the joint venture allowed the time and care needed to train Japanese craftsmen properly so things would not fall apart with the departure of foreign instructors. While the entire story of the America-Japan Sheet Glass Company occurs after the end of the Meiji period, as does the success of Asahi Glass, both companies were the result of an adaptive learning process that guided entrepreneurs, government investment and craftsmen of the Meiji period.

The success of domestic production was the defining factor that led to the diffusion of glass windows outside of government architecture. By the 1880s, glass doors and windows were common in public buildings and schools, because they were constructed along European models and required them. School students, used to studying in traditional structures, found the rooms "almost dazzlingly bright."[61] The use of windows in homes and shops came more slowly, not because of any cultural resistance but rather because only the rich could afford glass panes. Even as late as the 1890s, glass windows were still only seen in higher-end shops.[62] When Mitsukoshi Department Store opened in 1904, the show windows all had to be imported from Belgium. Mitsukoshi continued at the vanguard the following year by being the first to add glass showcases inside the store.[63]

Glass windows did not encounter much cultural resistance because they did not require any architectural change. It was not uncommon for a house with glass windows to be completely traditional in all other aspects of construction.[64] For example, when the Emperor moved from his temporary residence to a permanent palace in January of 1889, the new palace had glass shoji even though the rest of the structure was done in a very traditional Japanese style.[65] Even romantic purists like Tanizaki gave in to the need to have windows because he realized they were better insulators than paper shoji. Haiku poet Masaoka Shiki included mention of "The unfairness that sheet glass can't be made in Japan" in his list of ten things that were unfair. Other unfair things on his list included the unfairness of being unable to read postal cancellations, the failure of Westerners to appreciate sake, the unfortunate fact that country roads were never straight and that humans can't grow wings. Foreigners, not wealthy enough for Western houses, often had to adapt by ordering special shoji with glass panes. Alice Mabel Bacon, a teacher in the Peeress' school who lived in a house that had a Western wing built on for her use, specially ordered such screens with glass panes because she found that she had the choice of being either drafty or shutting the rain doors (*amado*), which would have required that lamps be lit constantly.[66] Real architectural changes would have been required to fit window sashes to slide windows up and down. The resistance to architectural change may explain why even today windows in Japan tend to slide back and forth rather than move up and down.

The kitchen was often the first room to have windows added because it was the darkest place in the house.[67] In the 1870s, American educator and scientist Edward Morse observed houses in Tokyo with glass panes in a row along the lower portion of the shoji screens in a way that allowed an unobstructed view outside to one sitting on the floor.[68] Glass windows began to appear in rural areas by the late 1880s[69] but even wealthy families did not have glass windowpanes on all of their outside doors and windows until the 1930s.[70] Glass pane usage was first higher in the north, where insulation and light were more important.[71]

Sheet glass production was a truly modern industry for the Japanese. It represented a break from past traditions, but it did not inherently represent

Westernization. One contemporary observer termed modernization in Japan to be a mechanical rather than a chemical process, two streams flowing side-by-side, like oil and wine, where each remains distinct.[72] With the exception of the rich who had Western-style houses or Westernized rooms attached to traditional houses, most people introduced glass into traditional lifestyles. Panes of glass in shoji screens, and other glass products such as wind chimes, sake cups and bottles were all present in early modern Japan, in the Meiji period they just became available to a wider public. Other glass products introduced into Japan, such as lamp chimneys and scientific equipment represented smaller technical advances but created greater real change in lifestyle.

We often talk about the miracle of Meiji industrialization, but no matter how much money government or private entrepreneurs poured into glass production, it was an industry characterized by "a series of failures and misfortunes."[73] Success was achieved by building on the failures of previous ventures and by introducing new technology that did not require workers to be so highly skilled. Until the late Meiji period, these workers were called craftsmen (*shokkō*); they came to be called engineers (*gikō*) because the process shifted from hand to machine. Like the silk industry, fortuitous timing helped the fledging factories take root. World production dropped due to the turmoil of World War I, which created a demand for what was initially lower quality Japanese glass. Nonetheless, supplementary industries like mirror production were still relying on imported glass as late as 1916.[74]

The high demand for investment in mechanical technology meant that sheet glass production was concentrated among a few companies, while other kinds of glass production, like bottles, occurred all over Japan. Today only three companies, two of which have been mentioned in this paper, produce all the sheet glass in Japan. (The third, Central Glass, was not founded until 1958.) While many important Meiji industries such as textile manufacturing are no longer significant, sheet glass is as much a source of trade friction today as it was in 1930, when Japan was first accused of dumping. The U.S.–Japan Flat Glass Agreement of 1995 has been deemed a failure and sheet glass remains a bone of contention.

Natsume Sōseki, one of the acknowledged giants of modern literature used glass windows as a metaphor for ambivalence toward Westernization in works such as *Sorekara* (1909). The main character, Daisuke, is a dissipated man of 30 who lives off his father. Daisuke reads Western literature and has glass doors in his house, a modern man. Daisuke explains to his school friend that, "it's because the relationship between Japan and the West is no good that I won't work." He continues, "A people so oppressed by the West have no mental leisure, they can't do anything worthwhile."[75] Yet, the very glass doors that symbolize Daisuke's sentiment were a product of Japan's successful struggle to throw off Western oppression, a miracle of very industriousness that Daisuke distained.

NOTES

1. *Bunmei kaika, fukoku kyōhei, ryōsai, kenbo.*
2. Susan B. Hanley, *Everyday Things in Premodern Japan* (Berkeley: University of California Press, 1997).
3. Tanizaki Jun'ichirō, *In Praise of Shadows*, trans. Thomas J. Harper and Edward G. Seidensticker (Stony Creek, Ct: Leete's Island Books, 1977), p. 18.
4. Donald Keene, trans., *Essays in Idleness—the Tsurezuragusa of Kenkō* (New York: Columbia University Press, 1967), p. 50.
5. Yoshinobu Ashihara, *The Hidden Order—Tokyo through the Twentieth Century*, trans. Lynne E. Riggs (Tokyo and New York: Kodansha International, 1986), p. 13.
6. James Jackson Jarves, *A Glimpse at the Art of Japan* (Philadelphia: Albert Saifer, 1970), p. 115.
7. Christopher Dresser, *Traditional Arts and Crafts of Japan* (New York: Dover, 1994), p. 247. For example, British Consul Rutherford Alcock also expresses this theory in *Capital of the Tycoon* (1863) and *Arts and Industries of Japan* (1878).
8. F.G. Notehelfer, *American Samurai: Captain L.L. Janes and Japan* (Princeton, N.J.: Princeton University Press, 1985), p. 169. In 1870, Janes was invited to Kumamoto to run a school.
9. Sakita Yōsuke, *Nihon garasu kagami hyakunenshi* (Osaka: Nihon Garasu Kagami Kōgyō, 1971), pp. 2–3.
10. Dorothy Blair, *A History of Glass in Japan* (Tokyo: Kodansha International, 1973), pp. 178–179.
11. Arthur E. Fowle, *Flat Glass* (Toledo, Ohio: The Libbey-Owns Sheet Glass Co., 1924), pp. 17, 31–32.
12. Tanizaki, *In Praise of Shadows*, p. 2. The essay was originally published as a two part serial in the December 1933 and January 1934 issues of *Keizai ōrai.*
13. Information provided by John van Sant and Graham Healey.
14. Edward R. Beauchamp, ed., *Schoolmaster to an Empire: Richard Henry Brunton in Meiji Japan, 1868–1876* (New York: Greenwood Press, 119), p. 113. Itō detached himself from the main group and went on a second visit alone with Brunton.
15. Sidney Devere Brown and Akiko Hirota, eds., *The Diary of Kido Takayoshi* (Tokyo: University of Tokyo Press, 1985), Vol. 2, p. 289.
16. See Kume Kunitake, *Bei-Ō kairan jikki*, 5 v. (Tokyo: Iwanami bunko, 1978), Vol. 2, pp. 156–159 (St. Helens in Manchester), p. 336 (Chances), p. 342 (Osler's in Birmingham); Vol. 3, pp. 192–195, 201–206 (Belgium); Vol. 4, p. 355 (Venice), trans. in English as *Iwakura Embassy 1871–1873: A True Account of the Ambassador Extraordinary and Plenipotentiary's Journey of Observation through the United States of American and Europe* 5 v. (Princeton, N.J.: Princeton University Press, 2002), Vol. 2, pp. 154–161 (Manchester), pp. 372–375 (Birmingham), Vol. 3, pp. 194–199 (Belgium), Vol. 4, pp. 355–358 (Venice).
17. Kume, *Bei-Ō kairan jikki*, Vol. 3, p. 197.
18. Yetaro Kinoshita, *The Past and Present of Japanese Commerce* (New York: Columbia University Press, 1902), pp. 93–94.
19. Sakita, *Nihon garasu kagami hyakunenshi*, p. 20. According to Sakita, the first imports were made by an Englishman. These two firms are listed in Yokohama kaikō shiryōukan, ed., *Zusetsu Yokohama gaikokujin iryūchi* (Yurindō, 1998), pp. 68, 73.
20. *The 100 Year History of Mitsui & Co., Ltd.* (Tokyo: privately printed, 1977), p. 27.

21. Kozo Yamamura, "General Trading Companies in Japan: Their Origins and Growth," in *Japanese Industrialization and its Social Consequences*, ed. Hugh Patrick (Berkeley: University of California Press, 1976), p. 169.
22. Sakita, *Nihon garasu kagami hyakunenshi*, p. 20. Another was Sano-gumi.
23. Supposedly they are descended from the aristocratic Fujiwara family. See Oland D. Russell, *The House of Mitsui* (Boston: Little, Brown and Company, 1939), pp. 24–61.
24. Ibid., p. 186.
25. Yamamura, "General Trading Companies in Japan," pp. 170–171.
26. Tsuda Mamichi, "On the Trade Balance," in *Meiroku Zassh: Journal of the Japanese Enlightenment*, ed. William R. Braisted (Cambridge, Mass.: Harvard University Press, 1976), p. 326.
27. William Eleroy Curtis, *The Yankees of the East* (New York: Stone & Kimball, 1896), Vol. 1, p. 149. The value was US$183,883.
28. *Flat Glass: Nihon no ita garasu* (Tokyo: Ita Garasu Kyōkai, 2001), p. 10.
29. Dai Nihon Yōgyō Kyōkai, ed., *Nihon kinsei yōgyōshi* (Tokyo, privately printed, 1916), Vol. 4, pp. 96–97.
30. Ibid., pp. 266–268.
31. Yōgyōkyōkai, ed., *Nihon kinsei yōgyōshi*, p. 95. Nariakira ordered a crucible from Shigaraki to attempt production.
32. *Nihon Ita Garasu goju nenshi* (Osaka: Nihon Ita Garasu Kabushiki Gaisha, 1968), p. 23.
33. Inoue Akiko, "The Early Development of the Glass Industry of Japan," in *The Development of the Japanese Glass Industry*, ed. Erich Pauer and Sakata Hironobu (Marburg: Marburger Japan-Reihe, 1995), p. 14.
34. Flint glass is what we generally call lead crystal. It was patented by George Ravenscroft in 1671 and early defects were remedied by 1676 by adding protoxide of lead.
35. Sources are conflicting as to whether sheet glass was ever produced there at all. Some, such as Tsuchiya Yoshio, attribute the first glass to Iwaki Tatsujirō, *Nihon no garasu* (Tokyo: Shikōsha, 1987), p. 192.
36. It actually went through four variations on the name which all translate fairly closely in English.
37. Sone Tatsuzō, "Nihon shorai no jutaku ni tsuite," in *Nihon kindai shisō taikei toshi/kenchiku* (Tokyo: Iwanami shoten, 1990), p. 337. Condor is famous as the designer of the Rokumeikan, His best-known building that still stands is probably the National Museum at Ueno.
38. *Gojū nenshi*, p. 24.
39. *Gojū nenshi*, pp. 24–25.
40. Tsuchiya Yoshio, *Nihon no garasu* (Shikōsha, 1987), p. 190.
41. Blair, *A History of Glass*, p. 289.
42. Ibid., p. 289.
43. Later, natural gas was used. The Siemens furnace was also important in the iron and steel industries. Although German by birth, Siemens immigrated to England and was also known as Sir Charles William Siemens.
44. *Yōgyōshi*, pp. 27–33.
45. The reading of this name is uncertain.
46. *Yōgyōshi*, pp. 23–25.
47. J. Morris, *Advance Japan: A Nation Thoroughly in Earnest* (London: W.H. Allen & Co., 1895), p. 386. Morris was employed by the Department of Public Works in Tokyo.

48. Blair, p. 290
49. Okuma Shigenobu, ed., *Fifty Years of New Japan* (London: Smith, Elder & Co., 1909), p. 600. The figures are for 1900.
50. *Yōgyōshi*, p. 96.
51. *Gojū nenshi*, p. 32.
52. *Yōgyōshi*, p. 266.
53. *Gojū nenshi*, pp. 34–35.
54. Information provided by Ōnaka Kentarō of Asahi Glass.
55. University of Toledo, Libbey-Owens-Ford Glass Company Records, Doc. No. 2UQGC, December 8, 1917, p. 160 ff.
56. Fowle, *Flat Glass*, pp. 47–57.
57. *Gojū nenshi*, pp. 41–43.
58. Doc. No. 2UGOC, pp. 160–163 ff.
59. *Gojū nenshi*, pp. 64, 77.
60. *Yōgyōshi*, pp. 111–112.
61. Tokutomi Kenjiro, *Footprints in the Snow*, trans. Kenneth Strong (New York: Pegasus, 1970), p. 110. Japanese title *Omoide no ki*. While this is a novel, it is semiautobiographical and the impressions can be regarded as honest.
62. Douglas Sladen, *The Japs at Home* (London: Hutchinson & Co., 1892), p. 17.
63. Nakamura keisuke, *Bunmei kaika to Meiji no sumai* (Tokyo: Rikōgakusha, 2000), p. 170; Edward Seidensticker, *Low City, High City* (New York: Alfred A. Knopf, 1983), p. 111.
64. Kondō Yutaka, *Meiji shokki no gi yōfū kenchiku no kenkyū* (Tokyo: Rikōgakusha, 1999), p. 223.
65. Ibid., p. 88 and Donald Keene, *Emperor of Japan* (New York: Columbia University Press, 2002), p. 420.
66. Alice Mabel Bacon, *A Japanese Interior* (New York: Houghton Mifflin and Company, 1893), p. 38. See also, e.g., Arthur Collins Maclay, *A Budget of Letters from Japan*, second edition (New York: A.C. Armstrong & Son, 1886), p. 154.
67. Shibusawa Keizō, ed., *Japanese Life and Culture in the Meiji Era*, trans. Charles S. Terry (Tokyo: Ōbunsha, 1958), p. 129.
68. Edward S. Morse, *Japanese Homes and Their Surroundings* (New York: Dover Publications, 1961), p. 132.
69. Shibusawa, *Japanese Life and Culture*, p. 130.
70. Ibid.
71. Inoue Akiko, *Garasu no hanashi* (Tokyo: Gihōdō, 1988), p. 50.
72. Augusta M. Campbell Davidson, *Present Day Japan* (Philadelphia: J.B. Lippincott Company, 1904), p. 6.
73. *Gojū nenshi*, p. 22.
74. *Yōgyōshi*, p. 84.
75. Translations from, Natsume Sōseki, *And Then*, trans. Norma Moore Field (Baton Rouge: Louisiana State University Press, 1978), p. 72. References to the glass doors appear throughout the book.

8

Modernity and Carpenters: *Daiku* Technique and Meiji Technocracy

Gregory Clancey

Carpenters are liminal figures in the history of technology, and in studies of architecture, labor, masculinity, and any other commonly recognized disciplinary cluster which might claim their story. The Anglo-American word "technology" and its Japanese translation *gijutsu*, both coined in the mid-nineteenth century to describe a new regime of continually reengineered devices and systems, were set more or less above (if not against) the existing world of wood and hand-tools.[1] Few figures seemed more outside this new rubric than carpenters, who by definition were tied to an age-old organic material, and assigned by necessity to small, widely dispersed work-groups difficult to subject to industrial management or discipline. In both Japan and the West, the image of the carpenter surrounded by shavings still immediately evokes for most people the world of the pre-"technology" past, even if he (and it is still typically a "he") now carries a union card, the hand-tool is electrified, and the lumber was ripped out of a rain forest by giant skidders.[2]

For these and other reasons carpenters are a "hard case" in the history of modern technological change. They can hardly be made to symbolize that process, yet there they are, not only present but also deeply engaged and even necessary at many turns. That most of the first few generations of factory-based machines in Japan and the United States were made largely out of wood—as were most of the factories themselves—points to a void in contemporary perceptions of what "technological change" looked, felt, sounded, and even smelled like to many who initially engaged it.[3] Our focus on "the cutting edge" (a phrase itself grounded in a world of hand-tools) tends to mask that intensified use of existing skills, resources, and forms which was often the most immediate response, and occasionally a sustained one, to calls for "industrial" transformation by states and capital.

I want to tell a story about Japanese carpenters which both locates them within the historical and discursive world of technology/*gijutsu* (rather than an ahistorical space of "craft"), but also records their sense—and the sense of others who took them up as objects—that they are not fully of that world; that their very existence, after a point in time, raised questions and problems about narratives of progress imagined and actively constructed by elements of the modernizing state and its spokespeople in the professions. My aim is not a history of carpenters, a project that is likely impossible from an empirical standpoint and of limited value from a theoretical one, but an historical reflection on the carpenter as simultaneously facilitator, threat, target, survivor, and eventually icon in a self-consciously industrializing Japan; the *modern Japanese carpenter* as a technological subaltern.

In Japan (as in the United States), building-carpenters remained one of the largest single groups of male tradesmen well into the twentieth century. They remain important, and well-organized, in the early twenty-first century as well. Their work was and is considered essential to economic development in both countries.[4] They are regularly presented to children in each culture as models of "good" and extremely useful noncollege-educated men who work outdoors with their bodies and hands. In a contemporary work-world transformed by Taylorism and Fordism, carpenters seem comfortingly skilled, multi-dexterous, self-directed, and physically mobile, not to mention masculine. Because lumber is culturally coded as "clean" (and "natural") those who handle it escape certain stigmas assigned to those who work closer to machinery and its liquids. Even environmental discourses about the overcutting of forests rarely assign guilt to carpenters' worksites, although most carpenters—and Japanese ones in particular—now stand at the very end-point of a process of global extraction with potentially vast environmental consequences.

There is also their historicity. At the typical suburban housing site in either Japan or the United States today, the people doing most of the actual building still call themselves carpenters/*daiku* despite successive attempts to coin less historically laden names, and use tools and work practices, which are in many cases obviously derivative from tools and methods in use prior to the coming of factories in both countries. The presence of electrical tools, and new materials such as plywood and sheetrock, competes with the continued ubiquity of hammers and nails (in the American case mostly) and handsaws and chisels (in the Japanese one). Carpenters not only still own their tools in many cases, but also wear them. The "tool-belt" is at the same time useful, symbolic, ornamental, and an expression of personal and group identity. Just as the present American 2 × 4 frame would be recognizable in many points to an American carpenter of the mid- to late nineteenth century, the typical Japanese *zairai-kōhō* house-frame and its joints would be understandable to a house-building *daiku* of the early Meiji period. And despite "globalization," and even the creation of an international lumber market, typical contemporary house-frames in each country are still largely alien in form and detail to carpenters in the other. Even the hidden parts—perhaps especially these—carry "culture."[5]

This is not a story about simple continuity, however, let alone convergence, but about change and survival in a maelstrom of shifting definitions. Behind the question "how have *daiku* survived?" is the more difficult "what has '*daiku*' meant?" or more specifically, who and what has defined the terms of change within this and other Japanese work-worlds? The frames "craft to industry," "feudalism to capitalism" or "traditional to modern" will not get us very far. Carpenters, too numerous to form an "exception" while too culturally determined to fit smoothly into a "case," complicate binary and unidirectional narrations of techno-cultural change. And they did so, we shall see, even in the Meiji era.

DAIKU

The translation of the Japanese word *daiku* into the English *carpenter* (and the subtle reinscription of the English meaning onto the Japanese word) was an act of Meiji-period linguistic reordering we need to explain rather than take for granted from the beginning. I therefore favor the Japanese term in the story that follows. *Dai-ku*—literally "great artisan"—had once applied to mastership in any art, but in the course of the Edo period (1600–1868) came to refer exclusively to those who made architecture from wood. *Daiku* were not a small artisanal elite; they may have constituted one-third of all artisans in Edo (Tokyo) in the last years of the shogunate, a population five times that of the next largest artisanal groups, *tatami-ya* (tatami makers) and *sakan* (plasterers), who were both dependent on *daiku* patronage and thus captive to the larger work-culture. The term *daiku* was further modified by prefixes according to specialty: *miya-daiku* built temples and shrines; *sukiya-daiku* the teahouses and residential buildings of the upper samurai; *ie-daiku* regular urban houses; and unprefixed village *daiku* oversaw what were largely communal building processes in the countryside. Some combination of sheer numbers, power within the building process (aided by guild organization), the power *of* the building process (the most resource- and labor-intensive creative spectacle of the Edo period), intricacy of technique (accompanied by an unusually wide range of tools), and a certain ceremonial mystique was doubtless what allowed Japanese woodworkers to monopolize the character *dai*.[6]

Japan was also unusual among urbanized Asian cultures—even forest-dwelling ones—in using wood so exclusively as a building material. Here is another key to the status of Edo-period *daiku*; stonemasons were virtually nonexistent as a rival occupational group.[7] The exclusivity of this Japanese relationship between making architecture and framing in wood was something alien to the culture of Europe—and to many parts of Asia in the same period, including China and Korea—where the dualism carpenter/mason described different technical systems and expert practitioners competing for primacy within the same building process. In these other cultures, moreover, the worker in wood was often the lesser of the two, in terms of status and sometimes—as in industrial Britain—in overall numbers, as masonry came to constitute the higher-order building system. Britain's industrial revolution occurred in a landscape largely denuded of forests (an ecological crisis which that "revolution" arguably arose to address). In Japan, as in the United States, seemingly unlimited forest resources and a huge, widely dispersed artisan class capable of working wood would prove crucial to the relatively cheap, quick, and flexible replication of European forms, be they machines or buildings. Japan's fabled "poverty of natural resources" is a high-technology discourse that overlooks this intensive exploitation of lumber.[8]

Daiku identity was not solely constituted by material, technique, or economic niche building. The Japanese construction process was also marked by successive religious ceremonies in which *daiku* assisted, or took on certain functions of, Shinto priests. Ceremonies for ground breaking (*jichinsai*), ridge raising (*jōtōshiki*), and to mark completion (*rakuseishiki*) all had European and American equivalents. But in late Edo and early Meiji Japan, building-site ceremony could be unusually elaborate and politically constitutive, requiring not only special costume but also a detailed knowledge of performative scripts. Possessing costume and script-knowledge was likely as essential to the identity *daiku tōryō* (master) as possessing woodworking tools, books, patrons, and so on. It helped measure their distance from (and authority over) other building-related artisans, who played only peripheral ceremonial roles.[9] It also acted out their closeness to Shinto and its priesthood, a religion deeply entwined with forests and trees. To this day Shinto shrines are nearly always wooden and *daiku*-built, while Buddhist temples, since the second quarter of the twentieth century, are allowably built in a range of materials, including ferroconcrete.[10]

Daiku as Ukeoi-shi (Contractors)

One of the first effects of the Meiji Restoration on *daiku*—and on artisan (*shokunin*) culture generally—was the abolition of urban guilds. The guilds' ability to regulate wages and police artisan mobility inside the cities and between city and countryside was officially ended by a series of decrees between Meiji 1 and 5 (1868–1872), although in larger centers such as Osaka, new associations of *tōryō* (masters) attempted to salvage some element of control. The lifting of *Han*-related guild restrictions on movement, however, likely had more effect on *daiku* than other classes of artisan, because the most important new construction sites were now largely outside the castle towns.[11]

The construction of the treaty port of Yokohama and new treaty-related sectors of older cities such as Kobe, Nagasaki, Hakodate, and Osaka—a project that actually began in the Bakumatsu but accelerated in early Meiji—had a large and permanent effect on the *daiku* world, and on building-construction more generally. The unprecedented size and nature of these public works projects summoned into being a new class of "contractor" (*ukeoi-shi*)—family-based companies, which by the late Meiji period, would dominate a newly national building market serving mainly the government and *zaibatsu* interests. Of today's "Big Five" Japanese construction companies, no less than three—Shimizu, Kajima, and Taisei (formerly Okura)—were first organized at the worksite that was Yokohama. A fourth—Takenaka—took its modern form in the equally intensive worksite of Meiji-period Kobe. Shimizu, which emerged as the largest Japanese construction company in the years prior to World War II ("the top of Mt. Fuji" according to its competitor, Takenaka Touemon), had projected its reputation nationally as early as 1868, when it built Tokyo's first hotel for foreigners in the new Tsukiji quarter. The contemporary Shimizu Corporation would capitalize on this origin story in the late 1990s by having its Space Systems Division work out plans for a "space hotel."[12]

Many of the "contractors" who emerged in the treaty ports were fully connected, at least initially, to the world of Tokugawa-period *daiku*. Shimizu-gumi was founded by an Edo *daiku* who had previously worked on the restoration of Nikkō and the repair of Edo castle. The founder of Kajima was another Edo *daiku* who had finished his apprenticeship in 1840, more than a decade before Perry's arrival. The Takenaka family were fourteenth generation *miya-daiku* (temple and shrine builders) in Nagoya when they took a contract to build army barracks, and then 24 brick warehouses in Kobe. Present-day Takenaka brochures illustrate wooden shrines, which the family designed and constructed in the Tokugawa period, and the company maintains a large carpentry-tools museum in Kobe as a further reminder of their artisanal lineage. Not all of the treaty-port "contractors" were *daiku*—the founders of what became today's giant Taisai and Obayashi construction companies for example, were well-connected merchants—and even firms founded by *daiku*, when they became large enough, ceased to do any labor in-house but subcontracted nearly everything to smaller *daiku* and other artisan (*shokunin*) families. In fact by the Taisho era (1912–1926), big general contractors like Takenaka Touemon, despite his family's *daiku* ancestry, would lead an assault on what he called "narrow-mindedness" in Japanese construction by importing American materials and methods (such as steel framing and reinforced concrete), and famous American cultural obsessions such as mechanization and efficiency.[13]

We can still trace elements of *daiku* culture, however, in *ukeoi-shi* organization. One was the continued responsibility taken for both design and construction. This system was likely given its present name of *sekkei/sekō* (design/build) in Japanese only when it became obvious that Westerners conceived these tasks as naturally bifurcated.[14] Japanese *ukeoi-shi* never became "general contractors" in the Anglo-American sense, in which "architecture" and "general contracting" were separately institutionalized. It remained commonsensical in *daiku* practice (and most other realms of Japanese artisan culture) that design and execution were a continuous act, even after university-trained architects emerged later in the Meiji period. In fact architecture departments at Japanese universities themselves—from early Meiji to the present day—would teach design and construction in a single balanced curriculum (and usually within faculties of engineering), in contrast to the norm in Europe and America, where design was decisively elevated over construction in the course of the nineteenth century using the slogans "art" and "professionalism" (not to mention "technology," which conveniently established a binary relation with "art"). University-trained Japanese architects would have often preferred—and sometimes tried—to establish a system more in line with that of Europe or America, in which architects in private practice monopolized design-work and construction companies simply built. But the companies countered by hiring large numbers of university-trained architects themselves, putting them to work in "design departments." The vast majority of design-work in modern Tokyo and Osaka, to this day, continues to be done either in-house or contracted for by construction companies such as Shimizu and Takenaka following the *sekkei-sekō* system.[15]

The contracting system may have preserved, rather than weakened, the existence of small *daiku* firms under the control of their own masters (*oyakata*). The big *ukeoi-shi* were, at least through World War II, mainly organizers of multiple tiers of subcontractors (*shita-ukeoi*), who retained their identity as *daiku* or *sakan* even as they were integrated into new types of projects. Although the big *ukeoi-shi* were clearly a force for technological change in building—eventually introducing concrete, steel, and other systems often in advance of architects—we know little about their influence on the organization and character of their more traditional subs; the extent to which subs were "captive" to large *ukeoi-shi*, for example, how (or if) they were encouraged to adopt new techniques and organization, or how many became *ukeoi-shi* themselves. We do know that they eschewed, after growing to a certain size, using salaried *daiku* or other skilled tradesmen as pacesetters for subs, despite Takenaka's fascination with general contracting in America, where such practices were normative.[16] In other words, functions traditional to *tōryō*—such as design on the one hand, and the coordination of artisans (now subs) on the other—seem to have been compartmentalized or bureaucratized as each company grew. Big contractors even continued to oversee a range of Shinto site ceremonies at which non-company architects (if involved at all in a company project) played distinctly subsidiary roles.

STATE *DAIKU (SAKUJIKATA)*

While some *daiku* made the transition to the rituals and tools of capitalism (a "capitalism" constructed partly by *daiku* rituals and tools) others became officials in the Meiji State. When Western-style architects began to be trained by the Ministry of Public Works (Kōbushō) in the 1870s, *daiku* were already holding positions of authority in the same ministry. The ministry's Eizen-ryō (Construction Bureau) was initially staffed by former *daiku* Asakura Seiichi and Tachikawa Tomokata, who had worked on the construction of the French-engineered Yokosuka Navy Yard and Iron Works under the Bakufu. The Bureau was headed from 1874 onward by Hayashi Tadahiro, another former *daiku* who had worked at both Yokosuka and Yokohama. Before architecture students began re-staffing ministerial *eizen* bureaus in the 1880s, Hayashi was designing and executing convincing "Western" government buildings in a neo-Palladian British colonial style, ultimately producing over 30 structures for the Public Works Ministry and five for the Navy, some in brick, but others framed in wood and then covered with stucco and given stone quoins (corners) to make them appear to be masonry.[17]

State *daiku* (called *sakujikata*) were often from old and high-status artisanal families. Tachikawa was sixth-generation, and Oshima Mitsumoto, who was initially in charge of construction at the Public Works Ministry's Tetsudō-ryō (Railroad Bureau),[18] was from the Koura, one of the oldest and most powerful of the court *daiku* families under the Shogunate. Knowledge of Western forms, techniques, and practices learned at sites such as Yokosuka and Yokohama was in their case added to inherited knowledge/practice, and did not necessarily supplant it. Another large group of State *daiku* worked in the Imperial Household Agency, housing the Emperor and Court. Members of both groups would mount challenges to the authority of university-trained architects into the 1890s.[19]

A major site of *sakujikata* initiative was Hokkaido. The colonization of the northern island began in earnest during the 1870s under the Kaitakushi (Colonization Ministry), which hired American civil engineers and "agriculture specialists" to work with Tokyo *daiku* like Adachi Yoshiyuki in constructing Sapporo and other new inland centers. The American advisors, being from a culture where carpentry was still more central to infrastructure-building than in Europe, expected from the beginning that Sapporo would be a largely wooden, New England-like city, dependent for its construction, expansion, and repair on carpenters and even sawmills. Hokkaido, like Yokohama, thus became a training-ground in Western (mostly American) design for *daiku* from various parts of Japan. As early as 1873, *daiku* working with American engineers had produced one of the largest buildings in the country, the prefectoral legislature, designed in the American state-capital style with an immense wooden dome. By the end of the decade, Adachi was making even large Western-style *faux*-masonry wooden buildings [such as the Sapporo Hoheikan of 1880] aided partly by American carpenter's pattern books, of which the Kaitakushi had a large collection. In the late 1870s and 1880s, as Kaitakushi *daiku* began migrating back to their home prefectures, Hokkaido-like "Western" buildings began to be erected all over Japan, especially in Tōkaidō (the northern part of the main island of Honshu) where many are still extant.[20]

WESTERNITY AS "COMPROMISE" OR PIDGIN

As Cherie Wendelken has noted, "Meiji-era construction was overwhelmingly executed by carpenters using methods and materials not unlike those used in the late Edo period."[21] The opportunity to experiment and develop skill with "Western" forms, however, seems to have been decisive to the success of those Meiji-period *daiku*—albeit a small and favored minority—who sought and found patronage with the state and the *zaibatsu*. In the 20 years between the opening of the first treaty ports in 1859 and the graduation of the first university-trained Japanese architects in 1879, ambitious *daiku* worked out forms and techniques which Japanese architectural historians now refer to as *wayō setchū* (Japanese–Western compromise) architecture.[22] This phenomenon encompassed private houses and the largest structures of the state. Foreign materials (including brick and stone) were handled and molded, hybrid forms worked out, and new programs satisfied all within the rubric of *daiku shoku* (*daiku* work).

Wayō-setchū was a form of pidgin, the equivalent of the other pidgins constructing and regulating the daily existence of Yokohama, Sapporo, and the other zones where Japanese and foreigners were intended to mingle. Compromise (*setchū*) occurred in different ways and at various places within and around each project, most obviously in forms and surfaces, where "Japanese" and "foreign" elements were created and mixed into appearances foreign (yet recognizable) to both cultures. Compromise was also possible, however, between the visible form/surface and the hidden form of the building's wooden frame. Even the most externally derivative *wayō* buildings often (although not invariably) had Japanese "guts." Wall surfaces had never been the primary carriers of meaning in *daiku* work, which traditionally lacked walls altogether. Surface as material and surface as meaning were in that sense coproduced with the coming of Westernity. On the other hand, the reservoirs of knowledge, practice, and ritual already pooled in wooden frames remained there for some time. The wooden frame was the techno-social "frame" for the process of building itself, the space within which Meiji-period *daiku* still organized and ritualized not only their own labor but also that of dependent artisans.

Wayō-setchū sometimes carried compromise into the wooden frame itself, however, evidencing a desire on the part of some *daiku* to not just please patrons by manipulating surface and ornament, but conduct real experiments with foreign building technologies. To take one example, the frame of a convincingly "Western-style" wooden house at Meiji Gakuin in Tokyo (the Imbry-kan), constructed for a foreign missionary named Imbry by *daiku* from Niigata in 1880, reveals an exceedingly complicated arrangement of Japanese, British, and American construction details. The roof of the Imbry-kan for example, shades from mainly Japanese parts to purely British ones. Yet the particular "British" choices seem carefully chosen for their compatibility with, if not similarity to, familiar Japanese framing members. The details of the outer wall system were all American (i.e., influenced by balloon framing) yet put together in a way no American carpenter had likely seen. If the Imbry-kan is taken as an extreme example of Euro-American influence extending even to the interior structure, it is still far from a textbook example of "technology transfer." It shows Japanese *daiku* making complex choices based on technical and cultural criteria we can hardly grasp at this distance, choices that likely varied with the *daiku* and even with the project.[23]

Centers of compromise such as Yokohama and Sapporo and the compromise neighborhoods (concessions) housing foreigners in Tokyo, Osaka, and existing cities, were produced for the most part by *daiku* working with foreign amateurs, such as merchants, missionaries, and even technical advisors (*oyatoi*) hired by the Tokugawa and later Meiji governments. The American *oyatoi* in Sapporo, for example, "experts" in civil engineering and agriculture, were amateurs in architectural design and construction. We know from surviving drawings in both Yokohama and Hokkaido that sketches drawn by foreigners were often redrawn and rendered more technical by *daiku*, their units of measurement being Japanese *shaku* rather than feet. Because *wayō setchū* was compromise between Japanese artisans and foreign amateurs, it did little to threaten a *daiku* sense of control or competence. Foreign merchants, missionaries, and even engineers were unable to impose their own technical authority, or "transfer" to *daiku* a detailed knowledge of Western carpentry, a specialized skill-set which in most instances they would have lacked. Like any Japanese client of a Japanese *daiku*, they brought to the worksite, at best, a schematic vision of what they wanted, and expected the *daiku* to work it out in reality.[24] It is likely that Western (and perhaps Japanese) clients helped illustrate their desires, in some instances, with the aid of Western architectural pattern books, such as those gathered by the Kaitakushi. Yet pattern-book literature, even in America, was designed only to aid the consultation process between artisan and client, and not to give the detailed technical instruction which the (Western) carpenter–reader was assumed to already have.

Although the ability to make *wayō* buildings was clearly sought, prized, and ultimately displayed by select *daiku*, we have little idea of what *wayō* actually meant to those who practiced it (the term itself was created only in the 1890s, when Japanese buildings were becoming increasingly "pure" imitations of foreign ones under the guidance of architects). In art–historical texts, *wayō setchū* buildings are sometimes displayed as naive "precursors" of *yōfū* (architect-designed Western-style buildings) and are severed, by this very act of naming, from a large corpus of shrines, stores, and houses, which continued to be made, often by the same people in the same period, with little or no "*yō*" (Western) character. The later are not only undisplayed, but also in most cases undisplayable, given that they were rarely drawn or photographed.[25] *Wayō setchū* does not seem to have produced separate classes of "Western-style" and "Japanese-style" *daiku*, except at the level of those large *ukeoi-shi* who served as the membrane connecting government and *zaibatsu* patrons with *daiku* subcontractors. Pidgins are languages that occur among languages, specifically for purposes of trade. Many *wayō* buildings of the 1850s–1890s were made specifically for foreign residency, to house the intermingling of foreigners and Japanese, or to house Japanese who were behaving in manners still coded as "foreign." Thus, they cannot be interpreted as prototypes for wholesale relandscaping in the manner of later student exercises in the Imperial University's architecture course.[26] Rather than being a replacement for something that already existed, or a "stage" in an "evolution" of architectural design, they seem to have represented an addition, a grafting, a supplement; a new set of forms that certain *daiku* had added to their repertoire without altering their identity, work organization, tools, or skills. In fact, the very ability of *daiku* to exactly replicate foreign forms was grounded in specifically Japanese technical competencies.

DAIKU GEOMETRY (*KIKU-JUTSU*) AND THE MAKING OF WESTERN FORMS

Daiku culture, like almost all artisanal work cultures, is most commonly rendered "understandable" in modern texts through its tools. Partly this reflects the materiality and survivability of tools as opposed to more fragile or intangible cultural signifiers. Museums exhibit tools, people collect them, and photographers convert them into attractive still-lifes for catalogues and magazines. While tools were undoubtedly important to the people who owned them—their very forms often testify to this—they can be fetishized in contemporary understandings beyond anything experienced by the artisan-owners themselves. The tool-display reinforces a modern sense of preindustrial work as "handy" and even simple—a hand-in-motion—rather than a whole body at work, with other bodies, or a mind at work among other minds, and in specific places and times. Sometimes this fits a pattern of deliberate submergence, related to the processes of domination and displacement—either capitalist, or colonial, or both—which caused the tools to be initially abandoned, and then collected, and finally displayed.[27]

While *daiku* tools have been subject to collection and display since at least the mid-nineteenth century, the extensive *daiku* literature of eighteenth and nineteenth centuries remained largely uncollected and unread into the late twentieth. *Daiku* books were not collectable as "art" because their illustrations were so technical, nor was their technical prose considered a "literature," which outsiders might enter. Through most of modern Japanese architectural history, *daiku* work-culture was thus interpreted as preliterate, a class-based construction that fit the story of an academic architecture displacing a habitual and largely oral "craft" practice. In fact, the publishing of technical literature was highly constitutive of late Edo-period *daiku* culture, and this publishing practice continued, indeed accelerated, in the early Meiji period.[28]

Turning to *daiku* literature can deepen our understanding of "compromise," or more specifically how certain *daiku* were able to extend their technical competence in the direction of Western forms, yet without necessarily adopting Western techniques. Nakatani Norihito has completed a detailed study of one type, *kiku-jutsu* books, mathematical treatises which explain the cutting and arrangement of timbers in shrine and temple construction, and helped to "standardize" Japanese building production in the course of the Edo period. Although *kiku-jutsu* books first appeared in the eighteenth century, they were actually written and published in greatest numbers after the Meiji restoration. Early Meiji *kiku-jutsu* consciously continued the tradition of Edo-period texts, which were based indirectly on the Japanese mathematical system known as *wasan*. Nakatani also points out, however, that *kiku-jutsu*, one of a number of mathematical systems for laying out carpentry in the Edo period, was the only one to survive the Meiji restoration. The reason, he argues, was the new value placed on flexibility. Being more "geometrical" than other existing systems, *kiku* could be used to lay out even unusual forms, such as "Western" roofs, as well as familiar Japanese ones. While the new popularity of *kikujutsu* in early Meiji had as much, if not more, to do with the freeing of guild restraints and the increased demands for building of all types—including "traditional" structures—by late Meiji the literature was being transformed by the demands of *wayō*, a word that increasingly appeared in the titles of the books themselves. By the 1890s actual images of Western carpentry forms, such as trusses began to appear in *kiku-jutsu* books with their layout and construction explained entirely in *kiku* graphics (i.e., in the same graphical form as the explanation of temple roofs).[29]

The *kiku* literature shows Meiji-period *daiku* turning to an existing technical dialect rather than a foreign technical language, but one which, within the range of existing choices, promised to best expand its speakers' abilities in the direction of new forms and objects. *Kiku* itself changed as trusses, domes, and other novel foreign forms passed into and through it. By the first decade of the twentieth century, it was incorporating and explained the geometrical systems of imported Euro-American carpentry books, capturing not only Western objects, but also Western explanatory systems. The term "capture" is more accurate in this instance than "merger," which suggests a certain symmetry or two-way readability. Western geometry, once incorporated and explained within *kiku*, was no longer "readable" as Western explanation. It is not even clear whether *kiku* explanations of Western geometry were readable by most Japanese architects, who were trained in systems of calculation identical to those practiced in Euro-America itself.[30]

As we shall see, however, by late Meiji the most self-consciously modernizing types of *kikujutsu-sho* were being written not by *daiku tōryō*, but by a small group of university-trained architects connected to government trade schools. In order to place this development in perspective, we need turn to *architecture*, a discipline separate from, competitive with, and ultimately transformative of the *daiku* world.

THE COMING OF ARCHITECTURE: *DAIKU* AS *CARPENTERS*

The compromising process I've described so far is one in which select groups, families, or individuals within *daiku* culture sought (or were given) opportunities to enhance rather than replace existing technical competencies. At least some *daiku* made the transition to the new roles of *ukeoi-shi* or *sakujikata*, and to the mastery of new Western forms within existing cultural rubrics. Their ability to experiment with new or expanded roles and practices came almost exclusively through state and *zaibatsu* patronage.

Parallel to this, however, was another process initially much less compromising; one in which organs of the Meiji State attempted to actively displace or reform the realm of knowledge/practice which *daiku* embodied and controlled. This less compromising reform sought to affect a much larger number of *daiku*, although its operation would be more gradual and never complete.

An instinct to assign *daiku* to a "feudal" past along with Buddhist priests and Chinese doctors is clearly detectable in the early years of the Meiji Restoration. Following the latest in an age-old series of fires in Edo/Tokyo in 1872, the Dajōkan (Council of State) declared that the entire city should eventually be rebuilt in brick and stone.[31] The more modest result, the district called The Ginza, was the first (and nearly the last) archeologically correct material and technical replication of a European urban landscape inside the Japanese capital; a purposely uncompromising techno-cultural space intended to transform Japanese building-construction by substituting the brick shell for the wooden frame.[32] The establishment of a formal, Western (i.e., British) architectural curriculum at the Ministry of Public Works' Kōbudaigakkō (sometimes translated as "College of Technology" or "College of Engineering") in 1877 institutionalized a similar instinct in the realm of education. An editorial in the first issue of the student-edited *Kenchiku Zasshi* (Architecture Journal) of 1888 declared that

> Gradually, we should make every building in Japan completely brick or stone. Academically trained people should be in charge of this, and architectural regulations should be set. This is the basis of a strong nation.[33]

Although this plan would be complicated by practical difficulties, philosophical uncertainties, and even contrary opinions within the architectural community over the next 100-odd years, it would never be wholly abandoned. It would often be defended on utilitarian, rather than cultural grounds, as better protecting cities and nation against fire, earthquake, and typhoon (although a counter-discourse on the unsuitability of masonry buildings for Japan developed as early as 1880, when The Ginza began to develop seismic cracks).[34] But it also crystallized the need among the first few generations of Japanese architects—shared with their counterparts in engineering, medicine, and so on—to replace existing artisanal control of production with a foreign-derived knowledge which might form the basis for a new professionalism. There were, by the end of the Meiji period, less than 200 Western-trained architects in all of Japan (most of them Imperial University graduates) while there were countless thousands of *daiku*.[35] How to establish a proper hierarchal relationship between each group and its knowledge claims was a central problem in Japanese architecture—as in arguably every Japanese academically based discipline with nonacademic competitors—through the end of the Meiji period and beyond.[36]

The importation of "architecture" via Kōbudaigakkō (and later the Imperial University), also meant importing the Western dualities *carpenter/mason* and *wood/stone*. "Architecture," as explained by European teachers and their textbooks, was mainly about stonemasonry, and an "architect," the expert-leader of the building process, was a descendant of medieval stonemasons. In the logic of this system, *daiku* were to be considered "carpenters," a trade that in Europe was very much dependant on masonry and general contracting. The term *daiku* was almost never translated into Meiji-period

English as "great artisan," as this would have confused the relationship between *daiku* and the new profession of *architect* [a word that derives, ironically, from the Greek "archos" (chief) and "tekton" (carpenter)].

Even in the Japanese, *daiku* was now increasingly defined as someone technically competent in a single material—wood—rather than one whose roles encompassed design, direction, construction, and ceremonial performance. "Architect" in Japanese was initially rendered *zōkagaku-shi* and then *kenchiku-ka*, new Meiji-period titles, which set themselves above *daiku* and its associations. Meiji period architects were hardly ever from *daiku* families. Like most Imperial university graduates, they were typically from the former samurai class and aspired to spend their lives in state service.[37] In other words, the coming of architecture meant the restriction of *daiku* to a new, and lesser, set of actions, definitions, and spaces ordered by prior European experience and the ongoing realities of Japanese class relations.[38]

Architecture's redefinition of *daiku* was carried forward on multiple overlapping fronts. One was the building-site itself. If the more expensive buildings were now to be of masonry, and architects were to be in charge of designing and erecting these, this required re-creating *daiku* as carpenters within a new building process. Carpentry was still necessary to "fill in" the shell of any masonry building—the roofs, floors, interior wall, ceilings, staircases, all needed to be framed in wood, even in European practice. Thus, *carpenters* were as essential to the performance of European architecture as *daiku* were a potential impediment. Making *daiku* into carpenters/tradesmen on the European model became a prominent goal of Japanese architecture in the Meiji period, and has arguably continued to be one through the present day, given that the process remains always incomplete and contested.[39]

Architecture simultaneously separated the *daiku* of the present from the *daiku* of the past—the designers of prominent temples, shrines, castles, and palaces, which from the Meiji period onward became constitutive of a new "national heritage"—through the medium of architectural history. Japanese architectural history, which emerged most decisively in the 1890s with the work of Imperial University professor of architecture Itō Chūta, was the cataloguing of ancient wooden *daiku*-designed buildings—often accompanied by measurement, and with an eye toward their preservation as cultural monuments. It also provided models for an intensified shrine building campaign accompanying the literal construction of State Shinto. Coinciding with a period of rising nationalism, the movement toward architectural history grounded the new and foreign discipline of Japanese architecture in a more Japanese past. It also coincided with a rise in seismic activity (and the emergence of the science of seismology), which cast doubt on the survivability of masonry architecture on Japanese ground. Thus did architecture co-opt *daiku* ancestors even as it attempted to reorder contemporary *daiku* under its own leadership.[40]

As Cherie Wendelken discovered, architecture in the mid-Meiji period enlisted at least one prominent living *daiku tōryō*, Kigo Kiyoyoshi, whose family had long been in service to the Imperial Court. Raised to public prominence by the construction of a new, partly wooden Imperial palace, Kigo was invited to lecture on *kiwariho* (the system of *daiku* proportioning) at the Imperial University in the late 1880s, and collaborated with Itō and others in major shrine building projects, the first and perhaps last time that university-trained architects would form themselves into an audience for a *miya-daiku*. That architects came to know the *daiku* past, through architectural history, in ways that *daiku* themselves could only access through architect-produced narratives, gradually gave the new profession confidence in its ascendancy as the natural inheritors of "Japanese architectural tradition."[41] It is no coincidence that this occurred in the very period *sakujikata* (state *daiku*) other than Kigo were being systematically replaced in the ministries by graduates of the new architecture schools, a process that Imperial university professor of architecture Tatsuno Kingo would describe in retrospect as "break[ing] the bad habits of the Edo period . . . the *sakujikata* habit."[42]

A third front of this architectural displacement of *daiku* knowledge was more technical: the development, within the Japanese academy, of competency with "Western carpentry" (*yōfu mokuzō kōzō*). Even as *daiku* work began to be celebrated in the West as among the world's great carpentry traditions—which according to Tatsuno, was one reason Kigo was invited to teach elements of it at the Imperial University—young Japanese architects became fascinated with "Western carpentry" (meaning mainly British roof-framing) as a technical realm which naturally complemented masonry wall-construction, and one which they alone might master. Japanese roof-construction—the very heart of *daiku* technique—was reframed by Japanese architects and their foreign teachers as historically interesting, but wasteful, overly complicated, unscientific, and even dangerous. The Western truss, whose stresses and strains were subject to exact calculation, and which allowed the introduction of elaborate screws, braces, and an array of metal fixtures alien to *daiku* practice, became a sort of icon of the new architectural regime; an object that *daiku*—at least those who wished to work under architect-supervised state patronage—would eventually be encouraged to learn how to build based on architects' descriptions. Although the increased seismicity of the 1880s and 1890s temporarily shook architectural faith in brick and stonemasonry, and set students to thinking more deeply about wooden architecture, this only intensified their desire to master truss calculation as the basis for a "wooden architecture" (*mokuzō kenchiku*) distinguished from *daiku* technique.[43]

DAIKU REEDUCATION

While Japanese architects began to speak in the late Meiji period in the name of a "Western carpentry" superior in many aspects to the work of *daiku*, and an architectural history which eventually drew clear lines between the *daiku* of the present and those of the past, architects remained an exceedingly small and concentrated professional group while *daiku* were as general as the act of building itself. If knowledge produced in the academy was to fundamentally reorder built Japan, architects had not only to imagine an "architectural world," but also people it with listeners, believers, and those who would take instruction. A social stratigraphy began to form with "architecture" at its apex, and much like traditional Japanese temple building, its construction was to proceed from the top down.

The Ministry of Education in the late Meiji period was intent on educating not just a technical elite at the Imperial University, but a much larger class of "technicians" or "technical assistants" (*gijutsusha or kōgyōsha*), who would operate under the leadership of engineers and architects more smoothly than the notoriously unruly artisan class (*shokunin*). Beginning in the decade of the 1890s, the foreign knowledge generated at the Imperial University was to filter out from the architecture course, via architect–teachers and textbooks, into a series of trade-related schools (*shokkō gakkō*) where a new type of *daiku* would be produced. Education—the peopling of an "architectural world" (*kenchiku sekai*) around the small core of university-trained architects—arguably became as much of a preoccupation of the Meiji architectural profession as the design of buildings or the supervision of construction. The first few generations of Japanese architects became, in effect, architect–educators, some taking jobs as trade-school principals and others writing "popular" manuals in which the technical details of Western architecture—including even Western carpentry—were to be explained.[44]

Vocational schools, wherever they appeared in the nineteenth and early twentieth centuries, were largely assaults on existing systems of apprenticeship—the method by which artisan culture reproduced itself generationally in Japan and elsewhere. In the United States, successive attempts to found vocational school systems from the late nineteenth century onward were often opposed (and in many cases successfully) by building trade unions, who saw them as producing interchangeable "workers" designed to "trade" on business' own terms. Meiji Japan lacked strong or extensive building trades unions, yet the State's attempt to capture apprentices away from masters proved no easier a task.

According to a census of 1882, nearly 30 percent of the artisan families in Tokyo were still practicing trades related to building construction.[45] The *shokunin* class as a whole was targeted for reeducation as early as 1881, when Education Minister Fukuoka Katei approved a plan for establishing *shokkō gakkō* (roughly "trade schools") in every prefecture. The word *shokkō*, a newly coined neologism, was meant to describe the new type of person such schools were to train. *Shokkō* was a combination of the first character from the word *shokunin*—the word for "artisan," and the character *kō* or *ku*, which, although formerly also related to artisanship (e.g., *dai-ku*) was increasingly coming to represent industry (e.g., *Kō-budaigakko* and *Kō-busho*). *Shokkō* evoked a modern working class.[46]

Tokyo Shokkō-gakkō, founded in 1882, was designed to train the first generation of teachers for the provincial schools. Graduates became either teachers themselves, or *shokkō-chō* (foremen) assisting *kōgaku-shi* (engineers, a category that included architects). Architecture was added relatively late to the curriculum, in 1902. As Shimizu Keiichi pointed out in his detailed study of this system, the government's priority was on constructing an export sector, so there was a certain logic in beginning with the training of factory or shop foremen. At the same time, however, the State may have considered the building trades particularly difficult to reform or capture, given their size, independence, dispersal, and local power.[47]

By the early 1890s, hesitancy had also developed within the Education Ministry itself about the wisdom of a frontal assault on apprenticeship, perhaps because of low enrollments in the government schools. The apprentice system was officially recognized by the Ministry in 1894, when indentured apprentices were allowed to practice what they had learned in school in their own shops, under the eye of their masters (*oyakata*). Whether this represented an authentic limitation on the State's ambitions, or simply a change in strategy, is difficult to know. But that the knowledge of masters was to be supplemented through formal schooling was still a critical intervention in the formerly private and patriarchal relationship of master and apprentice.[48]

The prefectoral *shokkō-gakkō* in Hiroshima was among the earliest to institute a *daiku* course, in 1897.[49] The nomenclature at Hiroshima and elsewhere tells us much about the new *daiku-shokkō* such schools were intending to create. First of all, *daiku* work was included within a new meta-category called *mokkō*, or literally "woodwork," which formed one-half of a binarism with *kinkō* (metal work).[50] *Mokkō* was actually a Nara-period word, combining the characters for "wood" and "work," which was revived in the Meiji period as a literal translation of the English "woodwork," a designation increasingly applied to industrial shopwork in Britain and America. Here was a decisive naming of the *daiku* as *carpenter*—someone identified above all with a particular material. Moreover, *mokkō* did not so much map on to *daiku* work as absorb it. The word came to cover everything from the framing of buildings to the making of small wooden bowls using turning lathes. Such a range of procedures, which were described in the pages of a single *mokkō* textbook, were beyond the practice or custom of any existing Japanese artisan. The purpose of the Hiroshima school, wrote its principal, was "to raise good *shokkō* who are appropriate for *mokkō* and *kinkō* in the future."[51] *Shokkō*, *mokkō*, and *kinkō* clearly evoked a future of flexible material-specific workers able to evolve away from traditional work-cultures and practices.

The *mokkō* course was sufficiently institutionalized by 1899 to merit a Ministry of Education textbook, *Futsū mokkō jutsu (Regular Wood-Working Technique)*. At first glance, the book appears to be a straightforward codification of what *daiku* actually did. The building–carpentry lessons begin with the proper use of common and traditional Japanese tools, and then proceed step-by-step through the assembly of a typical Japanese small house. The text is liberally sprinkled with small drawings of apprentices wielding saws and chisels, the framing of typical roofs, familiar joints, and so on. Given that the text actually constructs a Japanese house, and imagines the reader into that process, it might seem that tradition itself was being taught. The "Western carpentry" then fashionable at the Imperial University is represented in only one lesson—the very last—on trusses.[52]

Yet *Futsū mokkō jutsu* was not a simple codification of *daiku* technique or pedagogy. To begin with, the lesson on trusses, coming at the very end, orders what precedes it as elementary or preliminary. The rigorous exclusion of Western features or details from the basic lessons re-creates Japanese carpentry not only as a technically pure, culturally indigenous practice (bereft of *wayō setchū*), but also at the same time, a comparatively simple one. The truss lesson is the student's bridge to a second stage—a higher, more advanced form of carpentry that he will access once he has mastered the "regular" (the *Futsū* of the title). The future planned for *Japanese carpentry* by the Education Ministry was not obsolescence or eradication, but relocation within a new hierarchy of difficulty, importance, and value at whose pinnacle were the forms of "Western carpentry": for example, the ability to construct European-style roof-trusses for masonry buildings.[53]

Turning to the "traditional" content itself, the Ministry had done more than convert the verbal and nonverbal lessons of apprenticeship into a series of illustrated written exercises. The lessons are mostly about tools. The images in *Futsū mokkō jutsu* of boys holding saws and chisels were intended to illustrate the correct positions of hands and bodies. Much was made of practice, or motion; of training hand and body to move in an instinctive way. Indeed, during the first two years of training in *mokkō*, half the total instruction time each week was to be spent abstractly practicing the use of tools. There is a close convergence between the images in the Education Ministry textbook and American trade-school texts of the same period, which seek to recreate, through rote practice, "insincts" supposedly lost since the days of medieval craftsmanship (and often coincidentally useful to modern production regimes).[54]

The existing attitude toward motion and rote practice in the world of Meiji *daiku* is far from clear. Because *daiku* literature had no philosophical or even journalistic component, we can know little of what *daiku* in this period actually thought about how skills were best or most naturally acquired. Nor can we discount the possible collaborative role of *tōryō* in preparing the Ministry's textbook. Yet in framing *daiku*-work as mostly about tool-use, and education mostly about practicing repetitive motions, books such as the Education Ministry's *Futsū mokkō jutsu* created an essentially new type of "*daiku*" literature, quite different than the *kikujutsu-sho* or *hina-gata* produced by *daiku* themselves. Classic *daiku* literature, much like classic American carpentry literature, was about design and measuring. The correct method of wielding tools would have been learned quietly, gradually, and perhaps chaotically during apprenticeships, which were identical with male adolescence, and hence with daily life. The formal presentation of discrete actions in *Futsū mokkō jutsu* would have been as novel to Japanese *shokunin* as Diderot's *Encyclopedia* (whose overall form and purpose the Ministry's textbook resembles) would have been to eighteenth-century French artisans.

Futsū mokkō jutsu and the State's attempt to capture *daiku* apprentices seems to have been resisted. The book was withdrawn in 1904, only five years after its publication. The *daiku* courses in *shokkō gakkō* were eventually replaced with *kenchiku* (architecture) ones, where students were trained not to make buildings but to draw them. The *shokkō gakkō* were even upgraded in the same period to *kōgyō-gakkō*, or training schools for "lower-level engineers." The Meiji state seems to have abandoned this initial attempt to educate *daiku* apprentices, and turned instead to producing an army of young architectural draftsmen, foremen, and supervisors from a different pool of urban student.[55] Trade schools of the *shokkō gakkō* type would continue on the provincial and local levels, however, and a small but steady stream of school-trained *daiku* would continue to flow into the larger pool of apprentice-trained *daiku* until the postwar period, when a reinvigorated trade school movement would finally supplant the master-apprentice system almost entirely.

The state's switch to the production of "lower-level engineers" in *kōgyō-gakkō* did not mean its total withdrawal, however, from the project of *daiku* reform. The locus of effort seems to have shifted from the classroom to the realm of publishing. Educators in the state schools worked to reformulate *wayō setchū* into a pidgin which might be spoken between *daiku* and themselves, rather than between *daiku* and foreigners. The teachers, and especially the foremen and supervisors they trained, still needed to solve the problem of marrying the language of architectural plan-making—for example, truss calculation—with the *kiku* geometry widely understood in the *daiku* world, if only for purposes of communication at work-sites. Their medium became *kikujutsu* literature itself. Since the Edo period and into early/mid-Meiji, authorship of *kikujutsu-sho* had been the exclusive province of *daiku tōryō*. In the last two decades of Meiji, however, a reformed version of this literature began to be produced by teachers associated with the *kōgyō-gakkō* movement. *Kiku*, which as we saw earlier was already trending toward the absorption of Western forms, now moved even more decisively in that direction, though it never abandoned core techniques nor core emphasis on temple and shrine construction. In one sense, the architectural co-option of *kiku* was recognition by the academic world of the continuing power of the *tōryō* over his own workspace, including his apprentices, and the need to accommodate existing artisanal language and custom in the new practice of architecture. Yet it also signaled the end of a *tōryō*-produced technical literature.[56]

CONDITIONS OF SURVIVAL IN THE *DAIKU* SPHERE

Despite the trade school movement, the work culture defined by *daiku tōryō* and their apprentices continued to evolve with a certain degree of self-coherence outside the immediate purview of the state. In Japan as elsewhere, architects directly controlled only a fraction of the building world, mainly that portion under government patronage. The contracting companies maintained a greater sphere of influence, but their focus remained on large-scale projects increasingly executed in concrete rather than wood. *Daiku* who worked at sites superintended by architects or large construction companies (especially those who had attended *shokkō-gakkō*) indeed came to form a class of "tradesmen." They performed woodwork according to architect-drawn plans, although the actual cutting and layout of wooden forms continued to rely on *kikujutsu*. The vast bulk of Japanese construction, however, continued to be the houses, small stores, and institutional buildings such as temples and shrines, and this sea of largely wooden building activity continued to be dominated by *daiku tōryō* some steps removed from the world of architecture, trade schools, or big contracting. The Tokyo that burned following the 1922 earthquake was overwhelmingly a city of small *daiku*-framed wooden buildings, as was the Tokyo that burned in the 1945 air raids, and as was the city that was rebuilt during the American occupation of 1945–1953. Indeed, the amount of Japanese floor-space constructed through the medium of wooden framing was only exceeded by floor-space framed from other materials—mainly concrete and concrete block—in 1963, nearly a century after the Dajōkan (Council of State) had declared that Tokyo would become a city of brick and stone.[57]

Interviews with elderly *daiku* who remember the postwar reconstruction (and even, in some cases, the prewar work world) reveal a building process more notable for its technical and social continuities with the early Meiji period than with any model of technologically induced change. Mechanized saws, which had been common at the lumber-producing end of North American carpentry since at least the eighteenth century, and which had been made in small numbers by Japanese foundries since the early 1880s, only began to appear in Tokyo in the late 1930s, and remained uncommon until after the war. Electrical hand-tools, largely unknown through the end of World War II, gained ground only gradually through the 1950s. Until at least the Korean War, pointed out one Tokyo *tōryō* I interviewed, a constant supply of electricity at even a large building site could not be guaranteed, so no *daiku* could afford to show up without hand tools. The joints of the frame itself were gradually covered with metal fittings—earthquake-proofing devices mandated by law since at least the 1930s—but these merely supplemented and strengthened wooden joints underneath, which many *daiku* continued to believe—in contradistinction to the state and its architects—were inherently earthquake-resistant. Even the social organization of the typical postwar building site—an experienced *daiku tōryō* in charge of a crew of apprentices and journeymen, and ordering the work of other artisans around his own—was essentially preindustrial in character. All of this contributed to the *daiku*'s perception of himself by the postwar period, and even more so the perception of outsiders, as a figure who has preserved "tradition" within a sea of change.[58]

The construction of "the traditional carpenter," however, ignores the role of *daiku* in the making of the modern state and its economy, beginning with Yokohama and continuing through today's "housing starts." It particularly overlooks the willingness of Meiji period *daiku* to "compromise" with foreign knowledge—and bring existing knowledge to bear on the construction of foreign objects—a phenomena, which we could have continued to trace into the present day. On the other hand, the *daiku* world did not always evolve in directions chartered or controlled by state and *zaibatsu*, despite the *daiku* origins of Japan's modern construction firms. State-directed planning for *daiku* reeducation, which would resurface repeatedly through at least the 1960s, indicates that the status of *daiku* as "modern" and spontaneously contributory to the new Japan was always a fragile one, particularly in the eyes of their architect-rivals. While many Japanese modernizers of the Meiji period would have preferred that a nation of brick, steel, and "carpentry" spring into reality overnight, Japan's actual pageant of constant and sometimes rapid change would actually be endlessly coproduced from above and below.

NOTES

1. *Technology* and *gijutsu* are not actually identical in usage. The English word lends itself to drawing sharper distinctions between "modern" and "premodern" or "sophisticated" and "unsophisticated" skills, objects, and infrastructures than the Japanese one. *Gijutsu* is easier to use interchangeably with *technique*. There is, on the other hand, an arguably closer fit between *gi-jutsu* and *gi-shi*, the Japanese word for engineer, than there is between *engineer* and *technology* in English. For an introduction to the history of the English word *technology* see Leo Marx "The Idea of Technology and Post-Modern Pessimism," in *Does Technology Drive History?*, ed. M.R. Smith and Leo Marx (Cambridge, Mass.: MIT Press, 1994).
2. The continued, intensive use of lumber in Japanese building-construction, and Japan's heavy dependence on imported lumber since the 1960s, has directly contributed to the clear-cutting of rain forests from Indonesia to Canada. As of 1999, Japan imported 37% of all internationally traded wood products, a figure far in excess of that of any other country (although other heavy wood-users, like the United States, clear-cut their own forests). The largest single user of imported lumber in Japan was the domestic housing industry. Global Witness, "Timber Takeaway: Japan's Overconsumption: The Forgotten Campaign," Briefing Paper, Oct. 1999 <http://www.globalwitness.org/campaigns/ forests/japan/downloads/ takeaway.doc>
3. For a reflection on this paradox from the standpoint of American history, see the introductory essay, "The Experience of Early American Technology" in *Early American Technology: Making and Doing Things from the Colonial Era to 1850*, ed. Judith McGaw (Chapel Hill: University of North Carolina Press, 1994). Two scholars, one American and one Japanese, who have argued for a fuller consideration of carpentry in accounts of technology and architecture are Brooke Hindle, *America's Wooden Age* (Tarrytown, N.Y.: Sleepy Hollow Press, 1985) (reprint) and Muramatsu Teijiro especially his *Waga kuni daiku no kōsaku gijutsu ni kansuru kenkyū* (Tokyo: Rōdō Kagaku Kenkyūjo Shuppanbu, 1984). One of the few English-language accounts of Japanese technological change to acknowledge the important role of carpenters is Tessa Morris-Suzuki, *The Technological Transformation of Japan* (Cambridge, U.K.: Cambridge University Press, 1994), e.g., pp. 51, 90. Among Japanese surveys see Nakaoka Testurō, *Kindai Nihon no gijutsu to gijutsu seisaku* (Tokyo: Kokusai Rengō Daigaku/Tokyo Daigaku Shuppankai, 1986).

4. *Zenkensōren*, the largest federation of construction workers' unions in Japan, is made up largely of carpenters and plasterers who construct wooden houses. Its membership stood at 300,000 in 1990 (Sidney Levy, *Japanese Construction: An American Perspective* [New York: Van Norstrand Rheinhold, 1990]). The figure for the trade as a whole is certainly larger, as not all carpenters work in house-building. An informal estimate made in a trade publication of 1960, a period when wooden framing still accounted for the majority of new floor space created annually in Japan, suggested there were then more than half a million building tradesmen in the country who called themselves *daiku* (*Daiku kyōshitsu*, January 1960, 52).
5. In the late 1990s, around 80% of new Japanese wooden houses (which accounted for around 40% of all housing starts on an average annual basis) continued to be framed by *daiku*, often working for small firms, and using a post and beam (*zairai kōhō*) method unique to Japan. "Japan's Forest Industries: Analysis, Comment, and Forecasts" <http://www.blandon.co.uk/forestry/index.htm>
6. For discussion of the various building trades in this period see Hatsuda Tōru, *Shokunin-tachi no seiyō kenchiku* (Tokyo: Kōdansha, 1997). For *daiku*, see [in English] William H. Coaldrake, *The Way of the Carpenter* (New York: Weatherhill, 1990); and Kiyoshi Seike, *The Art of Japanese Joinery* (New York: Weatherhill, 1977), with a good introduction by translators Yuriko Yobuko and Rebecca M. Davis; and [in Japanese], Endō Motoo, *Nihon shokunin-shi no kenkyū V: kenchiku, kinkō shokunin shi wa* (Tokyo: Yūzankaku, 1961); Endō, *Shokunin-tachi no rekishi* (Tokyo: Shibundō, 1965), which covers a range of urban artisans, especially prior to Meiji; Muramatsu Teijiro, *Daiku dōgu no rekishi* (Tokyo: Iwanami Shinsho, 1973), the most detailed discussion in Japanese of *daiku* tools; and idem, *Waga kuni daiku no kosaku gijutsu ni kansuru kenkyū* (Tokyo: Rōdō Kagaku Kenkyūjo Shuppanbu, 1984).
7. Although stone-masonry had been practiced in Japan in antiquity, it survived into the Edo period mainly in the parabolic retaining walls for castle and palace moats, and as a tradition of stone bridge-making in Kyushu. Edo Japan was, overwhelmingly, a wooden place (*Kōdansha Encyclopedia of Japan* [Tokyo: Kōdansha, 1983], p. 170).
8. For a discussion of how central forestry resources were to the Edo period, see Conrad Totman, *The Green Archipelago* (Athens: Ohio University Press, 1989).
9. For *daiku* ritual, see Coaldrake; Endō, *Nihon shokuninshi no kenkyū.*
10. Sunami Takashi, "Architecture of Shinto Shrines" in *Architectural Japan*, ed. Japan Times and Mail (Tokyo, 1936), p. 13.
11. Suzuki Hiroyuki and Yamaguchi Hiroshi, *Kindai, gendai kenchiku shi (Shin kenchiku gaku taikei 5)* (Tokyo: Shōkokusha, 1993), pp. 244–245; Nihon Kenchiku Gakkai, ed., *Kindai Nihon kenchikugaku hattatsu shi* (Tokyo: Maruzen Kabushiki Gaisha, 1972), pp. 676–677.
12. For the early history of these firms see Kikuoka Tomoya, *Kensetsugyō o okoshita hitobito* (Tokyo: Shōkokusha, 1993). The quotation from Takenaka is on p. 79. For a contemporary account of these firms in English, see Sidney M. Levy, *Japan's Big Six: Inside Japan's Construction Industry* (New York: McGraw Hill, 1993). The sixth of Levy's "Big Six" is Kumagai-gumi; the other five have been collectively grouped in this manner since at least since the 1960s, so I've privileged the more historic and better-known "Big Five" in my own account.

13. Kikuoka; Muramatsu Teijirō, "History of the Building Design Dept. of Takenaka Kōmuten" in *Takenaka Komuten Sekkeibu*, ed. Building Design Dept. of Takenaka Kōmuten (Tokyo: Shinkenchiku-sha, 1987); idem., "The Japanese Construction Industry IV," *The Japan Architect*, January–February 1968 (318): 139–146; Takenaka Kōmuten Shichijūnenshi Hensan Iinkai, ed., *Takenaka Kōmuten shichijūnen-shi* (Tokyo: Takenaka Kōmuten, 1969).
14. Muramatsu believes that Takenaka's coining of the new word *Kōmuten* as its name for "firm" or "company" around 1909 sought to express the sense of a "design-build" entity (Muramatsu, "History of the Building Design Dept." p. 35).
15. Nihon Kenchiku Gakkai, *Kindai Nihon kenchikugaku hattatsu shi*, pp. 1939–1940; Muramatsu, "The Japanese Construction Industry IV," pp. 144–145.
16. Ibid.; Hazama notes that many Meiji-period "capitalist" firms, and not just the construction companies, made use of existing artisanal structures through subcontracting. There was also an appropriation of the term (and some aspects of the role) of *oyakata* (master) in new factory-based industries, as they made the transition toward foremen and wage-workers. Hazama Hiroshi, *The History of Labor Management in Japan* (London: Macmillan, 1997), especially chapters 2 and 3.
17. David B. Stewart. *The Making of a Modern Japanese Architecture* (Tokyo, New York: Kodansha, 1987), p. 31; Hatsuda, *Shokunin tachi no seiyō kenchiku*, pp. 206–208.
18. Kajima Iwakichi, one of the Yokohama *daiku* who founded a major contracting firm, did so much work for the railroad bureau that he was nick-named "*Tetsudō* (Railroad) Kajima" (Kikuoka, "History of the Building Design Dept.," p. 35).
19. Ibid.
20. Endō Akihisa, *Hokkaido jūtaku shi wa* (Tokyo: Sumai no Toshokan Shuppan Kyoku, 1994); idem., "Kaitakushi eizenjigyō no kenkyū," unpublished manuscript in author's possession, 1961; Koshino Takeshi, *Hokkaidō ni okeru shoki yōfu kenchiku no kenkyū* (Sapporo: Hokkaido Daigaku Tosho Kankakai, 1993). For an overview of the use of American experts in the colonization of Hokkaido see Fujita Fumiko, *Hokkaidō o kaitakushi Amerikajin* (Tokyo: Shinchōsha, 1993).
21. Cherie Wendelken, "The Tectonics of Japanese Style: Architect and Carpenter in the Late Meiji Period" in *Art Journal* Fall, 1996, *55* (3): 28–37.
22. There is a relatively large and growing literature, in English as well as Japanese, about the phenomenon of *wayō setchū* much of it illustrated with photographs and floorplans. The discussion below draws particularly on Fujimori Terunobu, *Nihon kindai kenchiku shi*, Vol. I; Suzuki Hiroyuki and Hatsuda Tōru, *Zumen de miru: Toshi kenchiku no Meiji* (Tokyo: Kashiwa Shobō, 1990); Hatsuda, *Shokunintachi no seiyō kenchiku*; Muramatsu, *Nihon kindai kenchiku no rekishi*, Chapters 1 and 2; Uchida Seizō, *Nippon no kindai jutaku* (Tokyo: Kashima Shuppankai, 1992), Chapter 1; Stewart, *The Making of a Modern Japanese Architecture*; and Dallas Finn, *Meiji Revisited* (New York: Weatherhill, 1995). The term "wayō," combining the characters for Japanese and Western, was common in the Japanese architectural world by the late Meiji period, judging from the number of style books that incorporate it into their titles. I use the term here in preference to a common synonym, *gi-yōfū* (imitation Western-style), which denigrates *daiku* efforts as derivative.
23. Gregory Clancey, *Meiji Gakuin senkyōshi-kan (Imbry-kan) no kōzō: Kenchiku chōsa hōkoku* (Tokyo: Meiji Gakuin, 1996).
24. Endō, "Kaitakushi."
25. A literature which symmetrically considers "wa" and "yō" in the work of Meiji period *daiku* awaits production. One step in that direction, however, is a master's

thesis by Yoshimoto Makiko, "Echigo maze *daiku*," unpublished M.A. thesis, Architecture, Hokkaido University, 1994, which follows Meiji-period *daiku* from Niigata prefecture as they move back and forth from Hokkaido to Niigata to Tokyo constructing *yōkan* (Western-style buildings) and traditional temples and shrines. A privately published history of a *daiku* family in Gifu prefecture, *Waza takumi hito Sakashita jinkichi* (Takayama, 1994) by Sakashita Yukari, discusses and illustrates the full corpus of their work from Edo to early Showa, demonstrating their movement from "wa" to "yō" depending on the commission.

26. In applying the linguistic concept of "pidgin" to the world of artifacts, I follow the discussion in Peter Galison, *Image and Logic* (Chicago: University of Chicago Press, 1997), pp. 48–51, 831–837.
27. For a nuanced discussion of collecting see James Clifford, *The Predicament of Culture: Twentieth Century Ethnography, Literature, and Art* (Cambridge, Mass.: Harvard University Press, 1988).
28. A group of architectural history graduate students at Waseda University, organized by Nakatani Norihito in the mid-1990s, was the first to systematically read and analyze *kikujutsu-sho* and *daiku* literature more generally. Their results were reported in Nakatani Norihito, "Bakumatsu, Meiji ki kikujutsu no tenkei katei no kenkyū," unpublished Ph.D. diss., Waseda University, Tokyo, March 1998, and Kurakata Shunsuke, "Meijiki kikujutsu no keishi tankai," unpublished M.A. thesis, Waseda University, Tokyo, 1994.
29. Nakatani, "Bakumatsu."
30. Ibid.
31. Muramatsu, *Nihon kindai kenchiku gijutsu shi*, p. 74.
32. For the construction of The Ginza see Muramatsu, *Nihon kindai kenchiku no rekishi* (Tokyo: Nihon Hōso Shuppan Kyōkai, 1977), Chapter 3; Fujimori Terunobu, *Meiji no Tōkyō keikaku* (Tokyo: Iwanami Shoten, 1982), Chapter 1; and idem., *Nihon no kindai kenchiku* (Tokyo: Iwanami Shoten, 1993), Vol. 1, Chapters 1 and 3; Stewart, Chapter 1; and Finn, Chapter 2.
33. Quoted in Muramatsu, "Mokuzō no kindaika," *in Nihon Kenchiku Gakkai*, p. 8.
34. For the Meiji debate over earthquakes and their effect on architecture and modern change, see Gregory Clancey, "Foreign Knowledge: Cultures of Western Science-Making in Meiji Japan," *Historia Scientiarum* March 2002, *11* (3): 245–260.
35. Muramatsu, "Nihon kenchiku gijutsu shi," p. 100.
36. This section summarizes arguments I make at greater length in Gregory Clancey, "Foreign Knowledge: Architecture, Seismology, Carpentry, Japan, and The West," unpublished Ph.D. diss., MIT, Cambridge, Mass., 1998, especially Chapter 1.
37. Ibid.
38. Ibid.
39. Ibid.
40. Clancey, "Foreign Knowledge"; Wendelken, "The Tectonics of Japanese Style."
41. Wendelken, "The Tectonics of Japanese Style," pp. 28–37.
42. Hatsuda quotes Tatsuno, one of the first Japanese architects to graduate from university in the 1880s, as saying this in a eulogy to another architect who entered ministerial service in the same period. "An honest wind blew into the architectural world," continued Tatsuno, in describing the transition from state *daiku* to state architects. Japanese architects, like their British and American counterparts, viewed the world of building artisans as essentially corrupt.

43. Ibid.
44. Meiji-period secondary education of building artisans and architectural "technicians" is dealt with at length in Shimizu Keiichi, *Meiji-ki ni okeru shotō chūtō kenchiku kyōiku no shiteki kenkyū*, unpublished Ph.D. diss., Nihon University, 1982; and Suzuki and Yamaguchi, *Kindai, gendai kenchiku shi*, pp. 271–279.
45. Suzuki and Yamaguchi, *Kindai, gendai kenchiku shi*, p. 271.
46. Shokkō would turn out to be a working-class elite, however, as the schools could not make real inroads against apprenticeship in the course of the Meiji era. The graduates came to specialize in government-related work, of which architects were increasingly in charge (Shimizu, *Meiji-ki ni okeru shotō chūtō kenchiku kyōiku no shiteki kenkyū*, pp. 13–18).
47. Ibid.
48. Suzuki and Yamaguchi, *Kindai, gendai kenchiku shi*, pp. 277–229.
49. Ibid., pp. 271–279.
50. Shimizu thinks that the "wood-metal" division had been institutionalized in these schools at least a decade before, in the later 1880s (Shimizu, *Meiji-ki ni okeru shotō chūtō kenchiku kyōiku no shiteki kenkyū*, p. 15).
51. Suzuki and Yamaguchi, *Kindai, gendai kenchiku shi*, p. 275.
52. Monbushō [Ministry of Education], *Futsū mokkō jutsu* (Tokyo, 1899).
53. Ibid.
54. Ibid.; According to historian Shimizu Keiichi, a volume entitled *How to Use Woodworking Tools*, produced in Boston in 1881 as a textbook by that city's own trade school movement, became the direct model for *Shokkō kyōto kasho*, a Japanese trade school text of 1888 that preceded the Ministry's official effort.
55. Shimizu, *Meiji-ki ni okeru shotō chūtō kenchiku kyōiku no shiteki kenkyū*.
56. For example, one of the most important and popular late Meiji *kikujutsu-sho*, *Nihon kenchiku kikujutsu-sho* (Tokyo, 1905), was written by Saitō Hejiro, who taught at a Tokyo *kōgyō gakkō*. Editions continued to be printed and sold into the 1930s. Nakatani, "Bakumatsu," pp. 85–87.
57. Muramatsu, "The Japanese Construction Industry IV."
58. This paragraph is based on a series of interviews I conducted with elderly and middle-aged *daiku*, including *tōryō* in Tokyo and Niigata in the mid-1990s. Yoshizaki Yōharu of Niigata, who ended his apprenticeship around 1952 and was given a set of hand-tools as a parting gift by his master, first saw an electrical tool on a worksite around 1955. It was a drill, used to help cut traditional mortise and tenon joints. A Mr. Narita, also from Niigata, did not use an electrical tool at a worksite until around 1960.

9

The Impact of the Great Depression: The Japan Spinners Association, 1927–1936

W. Miles Fletcher III

This chapter examines the impact of the Great Depression on business–state relations in Japan through a case study of the cotton textile industry. A common perception holds that the challenges of the Great Depression ushered in the era of state control of the Japanese economy. The experience of the cotton-spinning industry, however, suggests a different result. In fact, the challenges that executives overcame in the late 1920s and early 1930s bolstered their confidence in the efficacy of industrial self-governance and strengthened their conviction in the need for autonomy from the state. The most vexing issue became overseas trade barriers, a problem that first appeared in the 1920s.

In many ways, this analysis echoes themes raised in the previous three chapters by David Wittner, Martha Chaiklin, and Gregory Clancey. Just as Japanese in the late nineteenth century faced the abruptly imposed challenge of competing with Western imports and adjusting to Western industrial technology, spinning firms in the early 1930s had to confront a second set of foreign crises beyond their control in the form of both worldwide depression and the rise of trade barriers in crucial markets. If the Meiji government's economic policies in attempting to foster the silk and plate glass sectors had flaws, officials in the interwar period did little better in aiding the spinning industry. In fact, in the eyes of business leaders, the government's economic policies seemed often to do more harm than good. Spinning executives, as they had done before, took matters into their own hands and devised strategies for simultaneously regulating and expanding production with the aim of encouraging firms to modernize their equipment. The success of the spinning industry helps explain Japan's quick recovery from the Great Depression and its impressive expansion of exports and foreign trade by the mid-1930s.

The standard view of Japan's response to the Great Depression is that the collapse of the New York stock market in late 1929 coupled with the stringent fiscal policies of Finance Minister Inoue Junnosuke and the return to the gold standard in January 1930 dealt a devastating blow to the Japanese economy.

It remained in a quagmire until a new finance minister, Takahashi Korekiyo, decided in December 1931 to devalue the yen by abandoning the gold standard and to embark on a course of deficit spending. Although farmers suffered the most because of the abrupt loss of the huge American market for raw silk, the entire economy slowed for two years. As the prominent economic historian Nakamura Takafusa has argued, "The depression was especially severe in rural areas, where prices of rice and other agricultural crops fell 30–40 percent . . . These were wretched times in the urban areas as well, as labourers were laid off and merchants went bankrupt." "Industries were being ravaged by the recession across the board."[1]

Feeling acute pressure to respond, the government proposed and the Diet passed the Important Industries Control Law (IICL) in 1931 to promote the creation of cartels in vital sectors. Some observers contend that this experimentation with self-regulated cartels soon led to more heavy-handed controls,[2] while others attribute a more direct role to the Great Depression in the rise of the "managed economy" in Japan. Asserting that "crisis is the father of policy innovation," one scholar argues: "when the Great Depression occurred in the West, its influence, together with Inoue [Junnosuke]'s deflationary policy, drove Japan into the Shōwa crisis." In response to this trauma, the first stage of the development of a state-directed managed economy began in Japan as "a major departure from liberal capitalism," which had marked the Japanese economy beforehand.[3] It is tempting to discern a straight line from the IICL to economic controls enacted later in the decade for the purposes of war mobilization.

Yet, the Great Depression may have had a more complicated impact on Japan. As Nakamura himself points out, the effects of the Great Depression remain difficult to assess with precision. No accurate figures on unemployment exist, and "there were relatively few people without any work whatsoever." Moreover, "this recession was characterized not by a decline in production but rather by major price collapses and deficit-running operations in virtually all industries."[4] The nominal drop in the GNP, value of exports, and other economic indicators was striking,[5] but nominal figures did not tell the whole story. Although the nominal GNP declined by 10 percent in 1930 and 9 percent in 1931, the real GNP, a measure that takes deflation into account, actually inched upward by 1.1 percent and 0.4 percent respectively.[6] In the domestic context, the downturn in some ways proved less severe than that experienced in the recession that followed World War I. One of the foremost scholars of the depression era notes that the 36 percent slide in wholesale prices in Japan from 1929 to 1931 did not match the precipitous drop of 41 percent from March 1920 to April 1921.[7] Moreover, as Hara Akira observes, the "bottom of the period of the Great Depression came earliest in Japan, in 1930."[8] In these ways, the effects of the Great Depression in Japan differed significantly from the experience of the United States and European nations. The variety of terms that scholars use to refer to the period from 1929 to 1932 in Japan—the Shōwa slump, the Shōwa crisis, recession, and depression—reflect an uncertainty about the exact nature of what happened at that time.

To begin to clarify the impact of the Great Depression in Japan, this study examines the experience of one sector in that era, the cotton-spinning industry. It merits attention for several reasons. It became the first large-scale mechanized industry in Japan and the first industrial sector to become internationally competitive. In the 1920s, cotton textiles trailed only raw silk as the nation's leading export sector, and it accounted for roughly 16 percent of Japan's industrial production.[9] In brief, this study will argue that the Great Depression dealt a sharp but fleeting blow to this industry and in the process reinforced trends that were well under way. In fact, the sector was recovering before Takahashi had a chance to implement his bold fiscal and currency policies. The practices of this sector and its main trade association, the Japan Spinners Association (Dai Nihon Bōseki Rengōkai, abbreviated Bōren), did change in noticeable ways, but not in fundamental ones. Rather than mark a transition to a state-directed managed economy, this period instead reaffirmed the long-standing tradition within the cotton-spinning sector of self-governance. The main challenges facing Bōren derived from developments that predated the Great Depression: trade barriers in major markets, such as China and India, and competition from factories in China.

ADJUSTMENTS TO A NEW ERA

A brief survey of trends in the 1920s will help set the context for understanding the experience of the Japanese cotton spinning industry during the Great Depression by examining the growing challenge of trade barriers in vital markets, the role of Bōren, and its strategies for dealing with economic crises. The first seven years saw cotton-spinning companies confront rapidly changing circumstances on many fronts. In the previous decade, World War I had brought an unprecedented boom to the entire economy, as export markets in Asia, suddenly vacated by Western firms, opened up. The price of cotton yarn soared, as did profits. By 1918, the most successful spinning firms paid dividends of over 50 percent. A sharp postwar recession that began in May 1920 and lasted for over a year ushered in a period of much slower growth and lower profitability. The price of yarn became volatile. For example, in 1925 it ranged from 378 yen per bale to 274 yen.[10] The average dividend rate for spinning companies in the first half of that year fell to 16.9 percent from 21.1 percent in the first half of 1923 and 48.8 percent in the first half of 1920.[11]

Developments at home and abroad also complicated future prospects for the industry. In late 1923, the Diet passed a revised version of the 1911 Factory Law that moved up the established deadline for abolishing night shifts from 1931 to 1926. Because spinning mills customarily operated two ten-hour shifts for 28 days per month, the end of night work posed a major problem. So did the rise of nationalism in China, which absorbed 40 percent of Japan's cotton textile exports. In 1918, the Japanese government had acquiesced to a rise in China's import tariff, and in 1925 the multinational Beijing Tariff Conference considered the issue of granting tariff autonomy. In response, the spinners rapidly expanded production in

China during the 1920s in order to avoid the tariff, and they searched for new markets.

The view of Terada Tatsumi of the Fuji Gasu Spinning Company reflected the confidence of his peers in meeting these challenges. He knew that Japanese industry had to depend upon exports and processing raw materials from abroad. Industry had to "take the margin between raw materials and manufactured goods," and "this margin eventually is selling knowledge and labor." From this perspective, the cotton-spinning sector served as a prime example, because it had to import raw cotton from abroad—mostly from India and the United States—and export in order to pay for the raw materials. Japan had an advantage, because its work force was as skilled as those in Western nations but much less expensive. Through the use of technology, it could compete even in "high level" goods. The markets supplied by the British cotton textile sector, which boasted a staggering 60 million spindles, presented an inviting target. "If [the Japanese cotton industry] challenges British goods, the markets might be almost limitless."[12]

Yet, by the mid-1920s voices of concern about Japanese exports arose in the most promising of those options, India. Indian spinners complained about unfair competition in the form of the continuing practice of night shifts, credit allegedly granted to Japanese firms from the government, and official subsidies to Japanese shipping companies. A movement arose to demand higher tariffs on imported cotton goods. While Japanese diplomats in India worked quietly against this campaign, the Japan Spinners Association sent appeals to the Indian government and private business organizations refuting the accusations of unfair competition. When the Indian government in 1927 doubled the duty on thick and medium yarn under 40-count, Japanese exports dropped sharply but picked up in cotton cloth. By 1929–1930, the Japanese share of the Indian cloth market had doubled, from 14 percent in 1926–1927 to 30 percent.[13]

In China, political instability made the situation for Japanese firms and trade even more volatile. The Beijing Tariff negotiations broke down in 1926. The turmoil engendered by the conclusion of Chiang Kai-shek's Northern Expedition in early 1927 and his creation of the Nationalist Government in Nanjing prompted executives of Japanese spinning companies to appeal for government protection of their factories in Shanghai and other coastal cities.[14] Japanese forces did, in fact, intervene twice in 1927 and 1928. When anti-Japanese boycotts flared, Japanese exports suffered.

The need to eliminate night shifts created a huge problem. Since the 1890s, Japanese spinning mills had gained an advantage in overseas markets by running factories virtually around the clock to achieve a high rate of productivity per spindle. In late 1923, the devastation caused by the Kanto earthquake near Tokyo, which destroyed or damaged 20 percent of the cotton-spinning sector's total capacity, had resulted in a delay of the legislated end of night shifts from 1926 to June 30, 1929.[15] Outside pressures, however, guaranteed that the spinners would have to change their ways.

Indian business leaders and officials, in particular, stepped up their criticism of Japanese firms for violating the regulations of the recently formed International Labor Organization through the practice of night work and cited the reportedly poor working conditions in Japanese factories as a rationale for protectionist legislation.[16] Naturally, Japanese firms feared that the elimination of night shifts would reduce productive capacity and raise the per unit costs of manufacturing. To counter these trends, companies would have to undertake rapid expansion.

One distinguishing trait of the cotton-spinning sector remained constant: the dominant role of its trade association, the Japan Spinners Association. This group, which began in 1882, encompassed nearly all spinning firms and, since 1890, had implemented a policy of production cutbacks (*sōgyō tanshuku*, abbreviated *sōtan*) to deal with periods of reduced demand. Bōren had quickly become a major lobby for the industry and by the 1920s had even taken on a diplomatic role in presenting its case on trade issues to foreign business groups and officials.

The Japan Spinners Association early on established a practice of reaching decisions by consensus and relying on *sōtan* as the main method to adjust supply and demand. Annually the membership elected a governing Committee of 12 representatives from different firms. Because of a concern about the dominance of a small number of large firms, in 1923 an informal group, the Gosankai, began to represent the concerns of small companies. Afterward, two or three members from small enterprises regularly won spots on the Committee. It had the authority to draft appeals to the government or other interested parties and to create proposals for curtailments of production, but such measures could not take effect until approved by at least 90 percent of the membership. Members believed in *sōtan* as effective ways to enable firms to survive slumps in the market with productive capacity intact and ready to seize the next opportunity. The last cutback had occurred from May 1920 to December 1921.

THE PANIC OF 1927

As if coping with international trade friction in major markets and the elimination of night work were not enough, Bōren had to confront a major financial panic that hit the Japanese economy in 1927. As is well known, this crisis started with the collapse of several banks and the Suzuki Trading Company in March. The Diet's inability to pass a law to enable the Bank of Japan to issue bonds to the troubled banks made them more vulnerable. After a run on banks began, they declared a two-day holiday, and the Bank of Japan was finally able to step into the breach.

In April, the financial situation prompted Bōren and various cotton merchant associations to agree on a three- to four-week moratorium on payments for goods.[17] Declining sales of cotton yarn and anxiety about the general economic situation led the Japan Spinners Association to declare a six-month curtailment of production in April.[18] The *sōtan* stipulated that each company take four holidays per month—an increase of two holidays for most firms—and seal 15 percent of its spindles. After protests from two small firms that used much of their production of yarn for manufacturing their own cloth, the committee considered moderating the curtailment for such firms. In the end, the committee required these firms to seal just 8 percent of their spindles. As Nishikawa Hiroshi has pointed out, the *sōtan* affected small firms the most, because of the guideline of four holidays per month. While many larger firms already gave workers four holidays per month, small mills tended to grant only two days.[19] The head of the Japan Spinners Association, Abe Fusajirō, justified the action in terms of responding to the moratorium: rather than taking the "reckless act" of shutting down factories for three weeks or so, members should spread the decline in production over six months.[20]

If this action received only scattered criticism, the issue of extending and expanding the *sōtan* spurred controversy both within Bōren and outside of the group. The Committee first recommended extending the curtailment in July.[21] On September 20, the Gosankai, which represented smaller firms, called for a deeper cut in production, because they were caught in a squeeze between the falling price of yarn and the rising cost of raw cotton. Large firms that hoarded large stocks of cotton previously bought at cheap prices could still earn a profit of 15–20 yen on each bale, but smaller firms, which had to purchase cotton on the spot market, were losing 20 yen per bale. The views of large companies split on the proposal. Taniguchi Fusazō of Osaka Gōdō Spinning viewed further cuts in output as a justified act of "self-defense" when firms could not make a profit. Abe, who served as the president of the Tōyōbō company as well as the director of the Spinners Association, opposed expansion of the *sōtan*. He argued that the stagnation in the cotton industry was "worldwide" and that "there [were] fears that preventing exports further by raising the market price [of cotton textiles] in a time of slow exports would bring many large disadvantages to future development abroad." Probably referring to the public outcry over the rapid escalation of the price of rice as well as cotton goods in the late 1910s, he warned against additional mandated reductions in output "from the viewpoint of social policy."[22]

Meanwhile, other complaints emerged. At a meeting of the Gosankai on October 5, some companies opposed a *sōtan* because it would raise costs, hamper exports, and deflect attention away from the imperatives of creating new markets and improving productivity.[23] In addition, one outspoken leader in the Kantō area, Miyajima Seijirō of Nisshin Spinning, raised a difficult issue by demanding that Japanese firms in Shanghai, the location for most Japanese textile mills in China (the so-called *zaikabō*), enact a *sōtan*. Otherwise, domestic firms without factories in China would lose more of their market there.[24] The firms with factories in Shanghai countered that cutting output would only let Chinese competitors "run off with the prize." Instead of slashing output, domestic Japanese mills should shift production to finer yarns.[25] A squabble also arose over the idea of exempting yarn that Bōren's members themselves used to make cloth, as this measure obviously favored those with more extensive weaving operations.[26]

Outside of the association, enterprises involved with the export of cotton cloth and with the production of knitted goods protested the use of cutbacks to maintain a high price for cotton yarn. This type of criticism from other business groups represented a new development. In the past, some groups, such as associations of yarn traders, had pressured Bōren to enact cutbacks, but few had voiced discontent with such measures.[27] Producers of knitted goods now urged at the very least an exemption for yarn made into exports and, more radically, an end to the domestic tariff on imported yarn. In other words, they wanted to subject domestic spinners to outside competition, especially from Chinese factories.[28]

In response to this outside opposition and the lack of internal consensus, Bōren took the unusual step of appointing a special committee of representatives from 11 companies to carry out a snap study of the supply and demand for cotton yarn and cloth and to set the curtailment rate. This group promptly reported that stocks of cotton yarn remained at a level triple that of normal years and estimated that soon the surplus would grow to 5,000 bales per month. Therefore, the association would have to enact an increase in the curtailment rate in order to stabilize prices.[29]

By the end of October, the Spinners Association had settled on an increase of 8 percent for the *sōtan* with some special provisions that slashed the actual increase to 5.2 percent.[30] The total nominal rate would be 23 percent plus four holidays. The smallest enterprises with less than 30,000 spindles gained an exemption from the additional cutback as did the one member, Hattori Shōten, which used all of its yarn to produce its own cloth. The yarn that other members devoted to "private use" to weave cloth received a reduction of only 4 percent. Such measures reflected the long tradition of compromise within the association to achieve unanimity on, or at least acceptance of, the enactment of a curtailment. Two new provisions, however, stood out. First, an exemption for companies that eliminated night shifts provided a direct incentive to prepare for their legislated termination within two years. Second, the doubling of the curtailment rate of 8 percent for new spindles installed after November 1 indicated a major potential difficulty in the spinners' curtailment policies.[31] While various problems at home and abroad dampened sales of cotton yarn and cloth, the spinners had continued to add productive capacity at a brisk pace in order to prepare for the end of night work in 1929. Bōren's curtailments had always had the feature—perhaps distinct among cartel arrangements anywhere and often quite puzzling to outside observers—of never limiting the expansion of manufacturing equipment. Now, the nation's economic difficulties made the contradiction in this approach—that increases in capacity threatened to undermine attempts to restrain output—particularly obvious.

The curtailment, which the Spinners Association renewed every six months for the next two years, helped the spinning industry adapt to the challenge of ending night work. Continuing to offer an exemption from part of the cutback to firms that effected the change early provided an incentive to make the switch. Perhaps even more importantly, the cut in production helped to prevent firms that were striving to increase both manufacturing capacity and productivity from flooding the market and driving prices below the cost of production. The number of installed spindles increased by 764,000 (13.6 percent) between April 1927 and April 1929.[32] Moreover, executives took modest measures to increase productivity. The installation of small electric motors to drive spindles allowed the "separate operation of spinning machines" so that workers could fix broken threads by stopping just one piece of equipment instead of an entire section. Improving the selection of raw cotton and the increased use of high quality American cotton along with a move toward the production of higher counts of yarn meant less breakage and a faster speed for spindles. The number of women workers per 1,000 spindles dropped by 20 percent, and their total number by 11.7 percent. Overall, the costs for manufacturing 20-count yarn, the benchmark for the industry, fell by one-fifth, from 50 yen per bale to 40 yen between 1926 and 1929. Yet, the curtailments restrained output enough so that stocks of unsold yarn fell by about two-thirds between December 1927 and December 1928 and remained low.[33]

In his remarks to the annual conference of the Spinners Association in April 1929, Abe Fusajirō sounded cautiously optimistic. He observed that while the global cotton industry had been "inactive" for the past year, the Japanese sector had suffered less than some others because of the adroit adjustment of supply and demand through the *sōtan*. Citing the imminent end of night work, he described the current situation as a "time of change" and "one revolution in the spinning industry."[34]

Companies' results in the second half of 1929 presented little overt cause for concern.[35] Out of 52 firms that reported results, 42 paid dividends while another five registered profits but did not pay dividends. Two had gains for the period but carried over larger losses from the first half of the year. Only three companies had losses for both the second half and the entire year. Larger firms rewarded investors with the highest dividends as Kanebō paid out 35 percent and Fukushima Spinning 30 percent. Twenty-eight companies paid a dividend of 10 percent or less.

The economic strains of the late 1920s accentuated several problems that had been brewing for a while. One was the increase in new spinning firms that did not belong to Bōren, which together operated about 5 percent of the total spindles in the industry. This trend obviously threatened to undermine the effectiveness of production cutbacks, which had become continuous after the autumn of 1927. To Bōren's leaders, these outsiders also presented an unwelcome precedent for small firms in the association by "prov[ing] on the contrary the disadvantages of membership and inducing current members to leave." Observers agreed that the rebate that Bōren offered on the shipment of raw cotton from India to Japan, a bargain that had existed since 1893, kept the loyalty of many members. Unfortunately for the Spinners Association, some of the mavericks used mainly American cotton, and initial negotiations persuaded only one of five such companies to join. In response, the association struggled to find a way to apply pressure to the recalcitrant firms. They did not acquiesce until April 1930, after Bōren convinced six associations of cotton cloth and yarn merchants not to purchase cotton products from the renegades.[36]

The growth of the cotton-spinning industry in China, including Japanese companies, posed another challenge. Initially in the late 1920s, the spinners worried most about protecting their new factories in Shanghai and Qingtao against nationalistic protests by the Chinese and the confusion engendered by the Northern Expedition of the Nationalist Party as it unified China. As the situation calmed down, executives realized that the expansion of production in China would have effects beyond slashing exports of cotton yarn to that market. The lack of protective legislation and cheap wages gave mills there at least a 20 percent advantage in the cost of production over factories in Japan. Chinese yarn could threaten even the domestic market. By the start of 1928, Abe Fusajirō, as director of Bōren, asserted that because the worldwide recession had caused increased efficiency everywhere, the production of thick yarn and rough cloth should logically focus on the *zaikabō*, while domestic mills switched to finer yarn.[37]

In 1929, the government formed a special Council on Tariffs to consider changes in the levies on various products. As mentioned previously, certain sectors of the weaving industry, which was dominated by small enterprises, vocally advocated an end to the tariff on imported cotton yarn. The knitted goods sector, for example, had campaigned for such a revision since the mid-1920s. An internal report by Bōren clearly reflected the spinners' fears regarding such a measure. This study emphasized Japan's comparatively low tariff rates on cotton yarn and the modest profitability of spinning firms, whose average dividend rate had dropped to 7.9 percent. Without a tariff, cheap Chinese cotton yarn would "eat away at (*mushibamu*) every domestic market and endanger the base of the nation's industry" while impeding the reform of labor conditions and causing unemployment. Moreover, the benefits to weavers might prove transitory, as greater dependence on imports could make yarn prices more subject to conditions abroad and thus more volatile.[38] In a formal statement to Hamaguchi Osachi, the head of the Tariff Council, Bōren argued that the current tariff "protects fairness in trade" against the "great threat" of a Chinese industry whose plant exceeded one-half of Japan's productive capacity and which was unfettered by labor legislation.[39]

While making this case, the spinners showed little awareness of their own history and no sense of irony. After all, they were encountering in India similar complaints about "unfair" Japanese competition. The members of Bōren saw nothing wrong with Indian companies and consumers becoming dependent on Japanese imports. Nor evidently did the spinners recall that some 40 years earlier they had waged a fierce political campaign to overcome domestic agricultural opposition to ending the tariff on raw cotton in order to lower their own manufacturing costs. Obviously, a concern for survival overrode any thoughts about maintaining consistent principles in regard to foreign trade. Some of the more thoughtful executives must have wondered how long artificial barriers, such as tariffs, could keep competition from lower-cost producers in check.

In the early fall of 1929 the leaders of the Japan Spinners Association could look back with pride on their use of *sōtan* to handle smoothly the twin tasks of adjusting to the financial panic of 1927 and abolishing night shifts on schedule. Bōren also succeeded in maintaining its control of the industry by corralling pesky outsiders. Productive capacity had grown significantly, while prices had stayed sufficiently robust to prompt an end to curtailments in July. Issues related to foreign trade in the form of increased tariffs in India and the movement at home to eliminate tariffs on Chinese cotton yarn posed the most difficult problems. The spinners quickly realized the need to diversify markets and places of production. Although executives had some concerns about the Minseitō cabinet's decision to return Japan to the gold standard in January 1930, they accepted the move and prepared to adjust to it.

THE GREAT DEPRESSION HITS

The sudden drop in the American stock market in October 1929 initially caused few worries. As Finance Minister Inoue Junnosuke held fast to the goal of instituting the gold standard, firms trimmed their purchases of raw cotton in order to take advantage of the expected higher exchange rate for the yen and the lower price for imported cotton in the new year. Confidence in the market for cotton goods as "daily necessities" remained strong.[40]

By the end of the year, however, concerns had deepened because of the continuing rise in output and stagnating demand. The executive board (*kanjikai*) of the Gosankai recommended a cut in production of 10 percent. A special committee appointed by the Spinners Association reported in January that the last quarter of 1929 had witnessed a spike in stocks of unsold yarn and cloth, the equivalent of 58,000 bales of yarn. Companies' increased capacity would likely lead to higher monthly production while China's huge market for imports would contract because its silver-based currency would lose value to the gold-based yen. Bōren had to limit production once again.[41]

The idea of a *sōtan* provoked a round of especially virulent protests from other textile groups. In December 1929, the Osaka Export Knitted Goods Industrial Association accused the spinners of refusing to "recognize the suffering of cotton textile manufacturers who are in sister industries." Raising the price of yarn for cotton textiles, "which are a necessity, would threaten the life of the people" and hinder exports. In January, a meeting of cotton yarn and textile groups, which included five associations of weavers, warned against a *sōtan* and advocated an end to the tariff on imported yarn. They pointed out that by eliminating foreign competition the tariff made production cutbacks possible. They lamented that Bōren as just one organization of business leaders could make decisions about the price of cotton yarn that "hinders the stability of the people's life, invites the unemployment of textile and spinning workers, and prevents the expansion of the national economy." They called for laws to prevent this type of "rash act," the election of Diet members who would remove the tariff on yarn, and the formation of a new "alliance of cotton textile enterprises."[42]

In January, the determination of the Spinners Association to enact a *sōtan* for six months prompted an unexpected and unwelcome intrusion from government officials. While an ad hoc committee discussed the specific provisions of a curtailment, a sudden visit by the head of the Factory Section of the Osaka municipal government shocked executives, especially because he delivered an informal but clear threat to retract a one-hour extension of the workday for spinning mills permitted by the Factory Law. The head of the Social Bureau of the national Ministry of Home Affairs, which administered the Factory Law, then publicly urged the spinners to mandate holidays rather than sealing a percentage of spindles in order to minimize unemployment and to advance a policy of one holiday per week in accordance with the standards of the International Labor Organization (ILO).[43] The spinners responded by proposing an idling of 10 percent of spindles plus two extra holidays (for a total of four holidays per month). A company, however, could substitute an idling of 3.6 percent of spindles per holiday.[44] The Ministry of Home Affairs then approved the retention of the one-hour extension of the workday only for firms that opted for four holidays per month.[45] The Osaka Federation of Social Enterprises (Osaka Shakai Jigyō Renmei) complained to both Bōren and the Ministry of Home Affairs about the "unease" that possible unemployment would cause. Nishio Suehiro, a labor leader, lambasted the spinning companies for implementing a *sōtan* while some earned dividends of 20 or 30 percent.[46]

The first-half of 1930, however, turned out to be bracing for even the strongest of the spinning firms. Out of 51, only 21 paid dividends and nine registered profits while paying no dividends. Thirteen reported overall losses, and seven experienced losses for the six-month period that were covered by reserves. Kanebō and Fukushima Spinning led the pack with dividends of 28 and 25 percent respectively, but only eight firms paid 10 percent or higher.[47] Describing the year as one of "relatively many troubles," Abe Fusajirō gave a gloomy assessment to Bōren's members at the annual meeting in April. Aside from the impact of the return to the gold standard and the retrenchment in the government's domestic budget, the Council on Tariffs had, despite the spinners' protests, reduced the tariff on imported yarn by 35 percent. The Indian government was considering yet another raise in its tariff on cotton goods. Abe wondered whether the future would see the textile sector "enter a period of depression beyond today." Abe ended with a plea for "cooperation, unity, and strenuous effort."[48]

The new Indian tariff particularly irked the spinners. It not only raised the duty on all cotton goods from 11 to 15 percent, but also added another 5 percent to non-British imports plus another charge on unbleached cloth by weight. The total levy for exports of Japanese unbleached cloth would now range between 20 and 38 percent. Bōren's appeal to the Indian and British governments pointed out that this "discriminatory" and "prohibitive" tariff aimed not at protecting Indian spinners but at preserving the market share of British rivals. In addition, the contraction of Japan's total trade with India, which had doubled over the past 15 years, would hurt Indian consumers.[49]

After executives consulted with officials at the Ministry of Foreign Affairs, Ambassador Matsudaira Tsuneo in London lodged vigorous protests. Concerned that the new tariff clearly "violated the spirit of international cooperationism" and constituted "one big threat to world peace," Bōren worked with other associations to organize a large meeting of 600 executives in Osaka. It sent a telegram to officials in Japan, England, and India criticizing the proposed tariff as contradicting the spirit of the Japan–India Commercial Treaty and as hurting the interests of Japan as the largest consumer of Indian cotton.[50]

The spinners feared the spread of "the principle of special privilege" throughout the British Empire. Abe wrote that Japanese industrialists would have to cope with this situation, just as a decade earlier they had begun to invest in China and to upgrade their yarn exports in response to a higher tariff there. This time, companies would have to switch from unbleached to processed cloth. This strategy would bring them into direct competition with English mills. "Originally, we wanted an international division of labor. Having fallen into this situation, whether we like it or not, we cannot help waging hand-to-hand combat with the manufactured goods of Manchester." Observing that tariff wars were breaking out everywhere, Abe was ready to lead the charge by exhorting his troops to "advance with still more courage."[51] In keeping with Abe's declaration that "the development of new markets is an urgent problem for the future of our spinning industry," the Spinners Association and cotton merchant associations agreed to sponsor major trade missions to Russia and Africa.[52]

By May, the issue of expanding the *sōtan* had emerged once more. In what was becoming a set pattern of decisionmaking, the Gosankai first recommended a further reduction of output by 5 percent. Pointing to a drop of over 50 percent in profits for the industry, Abe predicted an "unprecedented period of difficulty" marked by the closing of factories, "voluntary cutbacks," and lower wages. Making Japanese goods cheaper, he contended, would not help in such a poor market. In early June, the association settled on an increase of 10 percent of idled spindles for a total reduction of 27.2 percent (a 20 percent idling of spindles plus two extra holidays).[53]

Once again, the measure engendered considerable controversy. The Ministry of Home Affairs continued to argue for increasing the number of holidays as opposed to idling spindles both as a means of minimizing unemployment and of improving working conditions. Sotomi Tetsujirō, the head of the Japan Knitted Goods Export Association, blasted the spinners for continuing to exempt themselves from the laws of the marketplace. As he put it, "making profits and having losses is typical of commerce in the world." Only the spinners expected different treatment. Instead, he argued, weak firms should be allowed to fail and within a year or so cheaper goods would bring a recovery and an expansion of markets.[54]

The woes of the industry continued. The drop of the price of 20-count below 100 yen per bale for the first time since 1915 moved Abe to declare "a completely unprecedented depression." One of the perennial top performers,

Kishiwada Spinning, implemented single shift operations at its main factory for the first time since 1900 while maintaining a policy of avoiding layoffs. As an alternative, other companies increased the number of holidays to as many as eight per month.[55] A number of firms reduced wages. Kishiwada cut the pay for employees by 20–30 percent and Nisshin Spinning by 8 percent. Kanebō's proposal in April to slash wages by 23 percent sparked strikes at 3 of its 36 mills.[56]

As firms' losses continued because of the depressed price for yarn, a proposal to increase the *sōtan* arose yet again. On August 20, the Gosankai suggested an increase of 8 percent to bring the total rate to 35.2 percent in order to limit total production to 190,000 bales per month.[57] Critics, such as Sotomi of the Knitted Goods Export Association, warned against hiking prices as competition in international markets heated up. In September, Bōren enacted a curtailment of 34.4 percent by adding two holidays per month.[58] Companies' balance sheets indicated that the second half of the year was even worse than the first half. Only 13 firms paid dividends, while 15 others showed a profit for the six-month period. Twenty-four registered losses.[59]

RECOVERY BEGINS

Significantly, the spinners did not look to the government for much help in their crisis. Instead, they relied on tested policies of cutbacks combined with a continuing emphasis on improving productivity. In the midst of economic distress in June 1930 Abe somehow found the optimism to announce, "From my perspective this is the bottom of the market and I do not think it will worsen."[60] Perhaps stung or worried by criticism from other textile groups, Bōren in July organized a meeting with representatives from associations of cloth manufacturers and yarn merchants in order to coordinate domestic policies. The resulting resolution issued just a vague call for the government to establish "effective and appropriate policies" to "rescue the general financial world." The delegates endorsed stricter self-regulation by suggesting that the Spinners Association should encourage "each related organization to establish control of the weaving sector and consider the regulation of supply and demand." Finally, all related business groups were to cooperate in opening new markets and in negotiations pertaining to tariff negotiations.[61]

Individual firms' gains in productivity yielded more concrete results. Beyond the spread of new techniques adopted in the late 1920s, companies invested in new equipment, such as "high draft" machinery that could more rapidly draw yarn through the various stages of spinning cotton yarn. Other machinery simplified the preparation of cotton fibers for spinning—opening and cleaning the fibers through a scutcher, disentangling and forming them into continuous strands (slivers), and drawing them into lightly twisted slivers (roving). New inventions that accelerated or shortened these processes could reduce the number of machines necessary to produce cotton yarn by 34 percent and the number of workers by 43 percent. One prominent executive claimed that he slashed costs for one of his factories by one-half through installing the simplex draft fly frame, which combined the preparation and spinning of cotton into one operation. The introduction of automatic looms in weaving meant that one worker could handle from 20 to 40 looms instead of the previous maximum of six.[62] The quest for efficiency also took the form of changes in management strategy, as larger firms had each factory focus on a narrow range of products.[63] Overall, the productivity per worker in cotton spinning in 1930 rose by 13.8 percent.[64]

By the spring of 1931, prospects had brightened considerably. Already in October 1930, 20-count had become profitable again, and stocks of cotton yarn in the Hanshin (Kansai) area had dwindled to their lowest level since 1924. By March, the relatively high price of 20-count twist was even prompting some cloth producers to import less expensive yarn from China. A speculative fever hit the market for advance purchases of yarn, while exports to Africa, the Mediterranean, and India picked up.[65] In March, Bōren decided to moderate the *sōtan* by one holiday, calculated as the equivalent of a 3.6 percent idling of spindles, from April to July. An additional resolution in April to introduce a flexible rate, with a 2 percent moderation in July followed by a possible further decrease of up to 5 percent in October brought sharp protests by yarn merchants. Because they had purchased yarn months in advance, they did not want sharp increases in supply to cause a sharp fall in the price of yarn. Finally, the spinners declared that they would stick with just a 2 percent moderation for the second half of the year.[66]

By June, the financial results for spinning companies had bounced back strongly from the second half of 1930. Thirty-three firms, nearly triple the number six months before, paid dividends, while 12 reported profits without sharing them with investors. Only four companies reported losses during both the previous six-month and 12-month period. That only eight firms paid dividends of 10 percent or higher served as a reminder of the mildness of the recovery.[67] Still, new plans for expansion of capacity, which had virtually vanished in 1930, reflected executives' new confidence. Firms wanted to add 300,000 spindles between August and December 1931.[68] Abe Fusajirō's annual address to the members of the Spinners Association in April sounded a note of pride and cautious hope. He observed that the spinning industry had "passed through a great test unseen in recent years." He attributed much of the sector's success to the *sōtan*: "through the cooperation of each member the association has fortunately planned the regulation of production and the control of business and has barely been able to escape these difficulties up to today." He concluded with the overall assessment that "during these [past] two-three months [the cotton textile industry] has somewhat followed a smooth course, but members cannot become relaxed."[69]

If increased productivity and profits bode well for the cotton-spinning sector by the late summer of 1931, two events abruptly changed the economic outlook. First, the eruption of war between Japanese and Chinese forces in Manchuria in mid-September, the Manchurian Incident, sparked new boycotts of Japanese goods in China and a sharp drop in exports to that important market. The Japanese, however, had endured periods of intense boycotts before. England's announcement in early October that it would abandon the gold standard was truly distressing.

This decision stunned the spinners. England, after all, had served as a pillar of the gold standard. Moreover, textile executives viewed the British as their main rivals. Immediately after England's decision, Abe proclaimed, "What is most feared is the change in the competitive power of English manufactured goods because of anti-Japanese [boycotts] in China and the weak pound." The extent to which the value of the pound sterling fell would determine how "painful" competition with British goods would become. Within a few weeks, some observers estimated that British cotton textile goods had gained a 20 percent advantage in price over Japanese products.[70] Tsuda Shingō, the president of the large Kanebō Company, argued that Japan had to abandon the gold standard to remain competitive in the international cotton textile market. Noting that Japan's cotton textile sector depended on "spreading manufactured goods widely in the world," Tsuda believed that "Japan's ability to develop [had] been checked by Britain's re-imposition of an embargo on gold."[71] The editorial staff of one prominent business magazine observed that in only a few weeks the gold standard had become the most urgent political and economic issue in Japan.[72]

As business opinion turned against the gold standard, the Minseitō cabinet felt mounting pressure. Finance Minister Inoue stoutly defended his policies, but the cabinet fell in December partially as the result of an internal split. The new prime minister and finance minister from the rival Seiyūkai—respectively, Inukai Tsuyoshi and Takahashi Korekiyo—immediately instituted an embargo on gold.

Despite the panic that the devaluation of the pound sterling had induced, spinning companies posted comparatively good results during the second half of 1931. Thirty-six firms paid dividends, and 14 others reported profits. No firms reported losses for the past six months or for the past year as a whole.[73] Just seven companies, though, distributed dividends of 10 percent or more. The spinning sector, which had registered overall losses of over three million yen in the second half of 1930, posted profits of over 21 million yen in the first half of 1931 and over 22 million yen in the second half.[74] Although the number of spindles in operation decreased slightly between September and December 1931, monthly production inched upward. The number of male workers decreased, but the number of female workers, who comprised over 75 percent of the labor force, increased.[75]

How severe was the Great Depression for the Japanese cotton-spinning sector? As is well known, the Japanese economy as a whole recovered quickly after the devaluation of the yen in late 1931. The cotton textile sector prospered too, as exports expanded rapidly because of both the 40 percent plunge in the value of the yen and the development of new markets. If Japan of all the industrial nations recovered most quickly from the Great Depression, this analysis suggests that the downturn was even shorter for the cotton-spinning sector. To be sure, events in 1930 staggered the industry. The number of spindles in operation dropped by about 11 percent and monthly production by 17 percent. The number of male workers fell by 29 percent and that of female laborers by approximately the same margin—a total loss of 46,709 workers.[76] However, one can even question the severity of this brief "depression" for the spinners. The rapid expansion of spindles in the late 1920s makes the reduction of operating spindles in 1930 seem less drastic. The decline in the number of female workers had begun with the emphasis on improving productivity in preparation for the end of night shifts. The large cut in the work force compared with the smaller drop in production suggests the extent of progress made in efficiency. As the firms' balance sheets attest, a recovery started early in 1931, well before the yen's devaluation. The number of operating spindles and monthly production increased as the year wore on, although by December those figures had still not yet returned to the levels of December 1929. The work force, especially female laborers, expanded at a modest rate as well.

RESULTS

Bōren's relative success in surviving the challenges of 1930 and 1931 reaffirmed its leadership's belief in the importance of the association's autonomous self-governance, which included the ability to enact production controls. After describing the spinners' achievements during the past 12 months as something of a miracle, Abe in December 1931 emphasized the need for cooperation in the entire cotton textile sector. Developments abroad—yet another tariff hike by India, "unprecedented boycotts" in China, the end of the British gold standard, and the "policy of the fierce recapture of Oriental markets by [British] manufactured goods"—had made 1931 in some ways the "worst year ever experienced." The production of yarn, however, had increased. If sales in old markets had declined, the overall decrease in exports was "not remarkable." "Unbending efforts to develop new markets" had even brought more exports of processed and high quality goods. Abe warned that Japan's embargo on gold would not bring immediate prosperity, as imported cotton would cost more and tariffs would continue to rise in overseas markets. The spinners would have to overcome such challenges by relying on "appropriate control by the [spinning] sector (*gyōkai*) and the cooperation of related sectors (*kankei gyōkai*)." "If [we] have the right policies," he concluded, "I believe that [we] will create the basis for future development and pass through [the year] in relative calm."[77]

The persistence of curtailments of production well after the start of the recovery sparked heated public debate about their usefulness. During 1932, for example, firms added 517,790 spindles, an impressive increase of 7.2 percent, and virtually all members of the Spinners Association showed a profit.[78] Still, a curtailment rate of 20 percent plus two extra holidays, a total cut in output of 27.2 percent, remained in effect. To dampen the fever for expansion, Bōren set the curtailment rate for newly installed spindles at triple the regular rate for the first three months of operation. Some critics began to wonder how the association could justify imposing limits on members' output during a period of prosperity. Had the policy of *sōtan* become simply a blatant means for increasing profits?

In mid-1934 Bōren maintained a curtailment rate of 18.8 percent (11.2 percent plus two extra holidays). Members, meanwhile, planned to expand production capacity by 660,000 spindles that year, a gain of roughly 8 percent over the total of 8,092,958 spindles at the end of 1933.[79] This situation prompted renewed criticism of the *sōtan*. Itō Takenosuke, the head of the Itō Chū Trading Company and a prominent figure among yarn merchants, pointed out that what he mockingly called the "10,000-year curtailment" had started four years ago and that two years had passed since the "worst period" for the industry. For all of the annoyance caused by foreign trade barriers, cotton cloth exports per month had doubled since 1929; the monthly output of cotton yarn had risen by 10 percent since then; and stocks of cotton yarn had sunk to minimal levels. "I have from the beginning been a glorifier of curtailments," Itō wrote, "but they must be flexible." He advocated that the current *sōtan* should end in six months.[80]

The next year, however, the Spinners Association raised the curtailment rate twice because of a low price for thick yarn below 20-count and uncertainty in the market. While Abe publicly recognized the need to deal with the growing public skepticism about production cutbacks by predicting that "as trends become more complicated, the problem of curtailments is sure to become more contentious," Bōren established a special committee to study the specific issue of *sōtan* in relation to firms' persistent and vigorous expansion of spindles.[81] The main defense of the current policy centered on the need to encourage firms to invest in new and more efficient equipment without flooding the market with surplus goods. As one unnamed executive of a large spinning company explained, the alternative was to "pile up old equipment and to repeat the failure of the British spinning industry."[82] Imamura Kusuo, the managing director of the DaiNippon Spinning Company, contended that "Japan's spinning industry [had] come to dominate the world" because of the current system of *sōtan*, which made possible a carefully managed cycle of production cutbacks and expansion of production. He predicted that in order to remain internationally competitive Japanese firms should replace five million old spindles with high draft models because these would cut costs by nearly one-half. This strategy was simply a "means of self-defense for spinning companies, because as the cotton industry trade war [became] more severe there [was] a need to lower costs."[83] Based on the study committee's recommendations, Bōren in October enacted substantially higher rates of curtailment for new spindles that would last for the first six years of operation.[84] The spinners, however, showed no signs of abandoning the use of *sōtan*. In fact, several months later the association tacitly admitted their permanent character by creating a standing Control Committee.[85]

The attitude of Bōren toward the Important Industry Control Law, which the Diet passed in March 1931, reflected its stubborn sense of independence and firm belief in the need for self-governance. In brief, the IICL permitted firms within designated industries to form cartels and, when two-thirds of companies in a sector agreed to form a cartel, it could request the Ministry of Commerce and Industry to force other firms to join. After nearly a half-century of independent operation, the leaders of the Japan Spinners Association obviously did not believe that they needed the government's help in forming or managing a cartel and they disliked the prospect of officials' meddling in their affairs. To their dismay, the law named cotton spinning as one of 24 industries that could form an officially sponsored cartel.[86] Executives were somewhat mollified by assurances from officials of respect for the principle of "autonomous control" by cartels and vows not to interfere with Bōren's activities as long as they did "not cause social problems."[87] Still, Bōren did not apply for approval under the law and remained a "self-governing organization" without formal legal status.[88]

The spinners wanted to maintain their distance from the Japanese state as much as possible. Hirasawa Teruo suggests that the threat of the Ministry of Commerce and Industry invoking the law probably made the Spinners Association more amenable to moderating production cutbacks after March 1931 in order to accommodate complaints from the weaving sector.[89] As discussed above, however, after 1930 curtailments with historically high rates of reduction in output became chronic, even after sales and profits improved dramatically. Furthermore, Miyajima Hideaki points out the hesitancy of the ministry to interfere with the success of the spinning sector after 1930 and officials' consistent rejection of pleas from the weaving industry to enforce changes in curtailments.[90]

Meanwhile, the spinners strived to extend their formal influence over the entire cotton textile sector. In October 1931 the governing Committee of Bōren met with three representatives of the Japan Federation of Cotton Yarn Merchant Associations to convince them not to import Chinese yarn "unless necessary."[91] In 1932, the Spinners Association sponsored a "Cotton Textile Conference" (Mengyō Kondankai) with yarn traders for the purpose of forming a broad consensus on the need to make periodic adjustments to curtailments of production. Even though this measure may have aimed, as some suspected, at "shift[ing] social responsibility for the *sōtan* to merchants" rather than heeding their views, the spinners took steps to create regular channels of discussion among different textile groups.[92]

In this regard, Bōren found that it had to contend with two new rivals, each sanctioned by specific legislation in the mid-1920s: industrial associations of weaving enterprises and export associations of merchants. Since the late 1920s, the Ministry of Commerce and Industry had encouraged cotton weavers to organize on a national level to prevent "excess competition" from lowering the quality of exports. By the early 1930s, the Japan Federation of Cotton Woven Goods Industrial Associations (Nihon Men Orimono Kōgyō Kumiai Rengōkai) had gained the legal authority to inspect members' products to ensure quality and to regulate production.[93] As noted above, Menkōren became a strident critic of the Spinners Association's use of curtailments of production to support a high price for cotton yarn.

In 1934, export associations, merchant groups that were organized by product and market, suddenly gained significant power in the wake of Japan's textile trade negotiations with India and the Dutch East Indies. These talks sought to pacify those governments' criticism of sharply rising Japanese exports. The resulting agreements required the strict regulation of cotton textile exports, as Japan was allowed in each case to export only a certain amount of such goods in return for predetermined purchases of imports. The Japanese government decided to grant export associations the power to decide which firms' goods would be exported. In general, the export associations distributed the production of 80 percent of the exports in each category according to the previous year's exports by each firm and opened 20 percent to competitive bids.[94]

Alarmed by the government's decision, Bōren first lodged vehement protests and argued for the creation of "export syndicates" controlled by producers. Meanwhile, Bōren and Menkōren each insisted that their members be able to participate in export associations. Moreover, the spinners and weavers put aside their rivalry to begin discussions in July 1935 on ways to gain back their influence over cotton textile exports. A month later, Bōren sponsored the establishment of a comprehensive organization to discuss issues related to the cotton textile sector, the Central Council for the Cotton Textile Industry (Mengyō Chūō Kyōgikai). It included representatives from both Menkōren and the Association of Cotton Yarn and Cloth Exporters (Yushutsu Menshifu Dōgyōkai).[95]

The leaders of Bōren worked hard to create a new structure of autonomous controls for the cotton textile industry. Meeting five times over the next nine months, the Central Council discussed matters related to foreign trade, the regulation of the manufacturing sector, and the control of exports.[96] In April 1936, the Spinners Association changed its regulations to permit the signing of official "agreements" with the Japan Cotton Association and the Japan Federation of Cotton Yarn Merchant Associations regarding the "buying and selling and the delivery of cotton yarn and cloth."[97]

Moreover, Bōren petitioned the government for reforms of the system of export controls so that producers could share the responsibility for decisions with export associations.[98] Accordingly, the spinners consulted closely with the most comprehensive organization of big business and probably the most influential business group, the Japan Economic Federation (Nihon Keizai Renmeikai), when it drafted a major proposal for a new system of national trade controls. It would empower a council of officials, producers, and merchants to determine general policies for exports and would guarantee through an arbitration procedure that export associations "respect the ideas of producers."[99] This lobbying had some effect, as the Ministry of Commerce and Industry in early 1937 submitted new trade legislation embodying such recommendations to the Diet, and it passed in August.

CONCLUSION

In brief, the Great Depression had a sharp but brief impact on the cotton spinning industry. The leadership of the Spinners Association described the situation as an unprecedented challenge. Indeed, a large majority of firms posted losses in 1930. By the spring of 1931, though, signs of recovery had already appeared. If England's abandonment of the gold standard—and, to a lesser degree, the Manchurian Incident—shocked executives that fall, their companies' balance sheets showed dramatic improvement by the end of the year.

The experience of handling a series of economic challenges from 1927 through the early 1930s strengthened the faith of spinning executives in the efficacy of their own self-governance and economic acuity. Carefully modulated cutbacks proved an effective economic tool. Although the sequence of nearly continuous *sōtan* started in 1927 as a response to a financial panic, the curtailments became a means of maintaining the price of yarn over the next two years while firms expanded rapidly to prepare for the end of night work in June 1929. Buoyed by the success of that strategy, the Spinners Association turned again to *sōtan* to adjust supply and demand in early 1930, after the worldwide depression began. Especially after England abandoned the gold standard in the fall of 1931, spinning executives publicly argued that Japan should follow suit and devalue the yen to make Japanese exports more competitive. After the market for yarn brightened, the curtailments remained, despite fierce criticism from various sectors, in order both to maintain prices and to encourage companies to raise productivity by installing more efficient machinery.

The spinners' experience during the era of the Great Depression did not strengthen the hand of governmental control in their sector. Indeed, the passage of the IICL in 1931 raised the specter of bureaucratic interference in the activities of the Spinners Association. Members, though, sought no direct aid from the government and emphasized measures that they themselves could take to overcome adversity—limiting production to maintain prices, increasing productivity, and developing new markets. Bōren declined to apply to become an official cartel approved by the Ministry of Commerce and Industry.

The main challenge to the cotton-spinning sector's prosperity came from trade disputes with major markets. Although the Great Depression aggravated this problem, it appeared well beforehand. In particular, criticism of surging Japanese exports of cotton goods to India in the early 1920s prompted the government there to impose a discriminatory tariff in 1927. In response, Bōren had to take a more active role in economic diplomacy. When diplomatic settlements of trade disputes after 1934 required limits on the shipment of cotton goods overseas, the spinners had to battle for control over the allocation of some goods for export. When the government assigned this task to export associations, the spinners resolved to organize the entire cotton textile sector and helped shape legislation that created a new system for the national regulation of trade and granted producers a major voice in decisions. Up to that point the Japan Spinners Association remained a strong, feisty, and independent-minded organization determined to govern its own affairs. Its encounter with the Great Depression only enhanced members' confidence in their ability to guide their industry through any crisis.

Not even Bōren, however, could cope with the demands of war mobilization caused by the China War and then the Pacific War. The China War, which began in July 1937, led within a few months to the introduction of direct governmental controls over trade. Officials strictly regulated the import of raw cotton because of the need for raw materials for munitions production. As the war continued and expanded into a conflict against the Anglo-American powers in 1941, the government introduced the rationing of civilian goods, and the production of textiles dropped precipitously. Bōren had to disband in 1942, when the industry came under the direct supervision of official "control associations." During the period of Allied occupation after Japan's defeat, though, the Spinners Association[100] reemerged in 1946 to guide the textile industry through a third major set of crises—that of wartime devastation and foreign occupation—and to help engineer the remarkable recovery of the Japanese textile industry by the early 1950s.

NOTES

1. Nakamura Takafusa, "The Japanese Economy in the Interwar Period: A Brief Summary," in *Japan and the World Depression: Then and Now*, ed. Ronald Dore and Radha Sinha (London: The MacMillan Press, 1987), pp. 52–67, on p. 61, and idem., *Lectures on Modern Japanese Economic History, 1926–1994* (Tokyo: LTCB International Library Foundation, 1994), p. 43.
2. Chalmers Johnson, *MITI and the Japanese Miracle: The Growth of Industrial Policy, 1925–1975* (Stanford: Stanford University Press, 1982), pp. 109–113.
3. Bai Gao, *Economic Ideology and Japanese Industrial Policy* (Cambridge: Cambridge University Press, 1997), pp. 72 and 74.
4. Nakamura, "The Japanese Economy," pp. 61–62.
5. Gao, *Economic Ideology*, p. 72. Although the author does not state whether the cited figures for GNP and exports are nominal or real, they approximate the nominal figures given in Dick K. Nanto and Shinji Takagi, "Korekiyo Takahashi and Japan's Recovery from the Great Depression," *The American Economic Review* 1985, *75*: 369–374, pp. 369 and 371.
6. Nanto and Takagi, "Korekiyo Takahashi," p. 369.
7. Hashimoto Jurō, *Daikyōkōki no Nihon shihonshugi* (Tokyo: Tokyo Daigaku Shuppankai, 1984), p. 165.
8. Hara Akira, "Keiki junkan," in *Nihon teikokushugi shi, 2, Sekai daikyōkōki*, ed. Oishi Ka'ichirō (Tokyo: Tokyo Daigaku Shuppankai, 1987), pp. 367–410, on p. 403.
9. I calculated this figure from data given in Takamura Naosuke, "Shihon chikuseki, 2, keikōgyō," in *Nihon teikokushugi, 2, Sekai daikyōkōki*, ed. Ōishi Ka'ichirō (Tokyo: Tokyo Daigaku Shuppankai, 1987), pp. 173–210, on p. 173.
10. Shōji Otokichi, *Bōseki sōgyō tanshuku shi* (Osaka: Nihon Mengyō Kurabu, 1930), pp. 461–462.
11. "Bōseki kaisha no gyōseki," *Chūgai shōgyō shinpō*, December 1, 1924 and "Waga bōseki jigyō no seiseki," *Chūgai shōgyō shinpō*, October 9, 1925, "Shinbun kiji bunkō: Menshi bōsekigyō," Vol. 12. These materials are housed at the Keizai Keiei kenkyūjo at the University of Kōbe in Kōbe, Japan.
12. "Waga sangyō ni okeru menshigyō no chi'i," *Chūgai shōgyō shinbun*, September 30, 1924, "Shinbun kiji," Vol. 12.
13. Nishikawa Hiroshi, *Nihon teikokushugi to mengyō* (Tokyo: Mineruva Shobō, 1987), pp. 274–303.

14. "ZaiShi bōseki wa kōjō heisa ga tokusaku," *Osaka asahi shinbun*, April 3, 1927; "Shanhai no kikki to bōseki no taido," April 9, 1927, *Osaka asahi shinbun*; "ZaiShi bōseki fuan o daku," March 31, 1927, *Osaka asahi shinbun*, "Shinbun kiji," Vol. 13.
15. Shōji, *Bōseki sōgyō*, pp. 455–458.
16. "Bōseki no shinyagyō teppai kiun o sokushin," *Chūgai shōgyō shinpō*, September 22, 1927, "Shinbun kiji," Vol. 14.
17. "Moratoriumu to mengyōsha no ukewata," *Kōbe shinbun*, April 28, 1927, "Shinbun kiji," Vol. 13.
18. Shōji, *Bōseki sōgyō*, pp. 467–473.
19. Nishikawa, *Nihon teikokushugi*, p. 179.
20. "Bōseki sōtan no ben," *Osaka asahi shinbun*, April 27, 1927, "Shinbun kiji," Vol. 13.
21. Shōji, *Bōseki sōgyō*, pp. 474–475.
22. "Bōseki sōtan kakuchō giron wa niba ni wakeru," *Osaka jiji shinpō*, September 24, 1927, "Shinbun kiji," Vol. 14. In 1918 a wave of "rice riots" in hundreds of cities, towns, and villages had shocked government officials. In 1919, the sharply rising price of cotton yarn caused officials for the first time to try to regulate its export.
23. "Bōseki jigyō no keiei gōrika no kyōgi," *Osaka asahi shinbun*, October 6, 1927, and "Bōseki sōtan kakuchō ni omomuita?" *Osaka jiji shinpō*, October 16, 1927, "Shinbun kiji," Vol. 14.
24. "Shanhai hōjinbō no sōtan jisshi ka," *Osaka mainichi shinbun*, October 26, 1927, "Shinbun kiji," Vol. 14.
25. "Sōtan kakuchō ketsuretsu ka," *Osaka asahi shinbun*, October 28, 1927, "Shinbun kiji," Vol. 14.
26. "Kiki o haramu sōtan kakuchō," *Osaka asahi shinbun*, October 27, 1927, "Shinbun kiji," Vol. 14.
27. W. Miles Fletcher III, "Co-operation and Competition in the Rise of the Japanese Cotton Spinning Industry, 1890–1926," *Asia Pacific Business Review* 1998, *5*: 45–70.
28. "Kanzei ni benpō o kōze yo," *Osaka asahi shinbun*, October 28, 1927, "Shinbun kiji," Vol. 14.
29. "Sōgyō tanshuku keika," *DaiNihon bōseki rengōkai geppō* 1929 (443): 25–32, on pp. 26–28.
30. "Sōtan zōritsu happun," *Osaka asahi shinbun*, October 25, 1927, "Shinbun kiji," Vol. 14.
31. "Bōseki sōtan kakuchō kakutei," *Osaka jiji shinpō*, October 30, 1927, "Shinbun kiji," Vol. 14.
32. I calculated this percentage from data given in Nishikawa, *Nihon teikokushugi*, pp. 179–181.
33. Nishikawa, *Nihon teikokushugi*, pp. 155 and 179–181; DaiNihon Bōseki Rengōkai Chōsabu, *Menshi bōseki jijō sankōsho, 1928 kahanki*, no. 52 (Osaka: DaiNihon Bōseki Rengōkai, 1929), p. 36; DaiNihon Bōseki Rengōkai Chōsabu, *Menshi bōseki jijō sankōsho, 1929 jōhanki*, no. 53 (Osaka: DaiNihon Bōseki Rengōkai, 1929), p. 36; DaiNihon Bōseki Rengōkai Chōsabu, *Menshi bōseki jijō sankōsho, 1929 kahanki*, no. 54 (Osaka: DaiNihon Bōseki Rengōkai, 1930), p. 36; "Nōritsu zōshin wa kenchō," *Chūgai shōgyō shinpō*, October 9, 1928, "Shinbun kiji," Vol. 15.
34. "Bōseki rengōkai," *Osaka asahi shinbun*, April 25, 1929, "Shinbun kiji," Vol. 15.
35. "Sakunen kahanki kaku bōseki kaisha eigyō seiseki," *DaiNihon Bōseki Rengōkai geppō* 1930 (450): 39–44.

36. "Mikamei kaisha ni Imen o uru nakare," *Osaka asahi shinbun*, November 8, 1927; "Ryō mosu kaisha kamei ka," *Osaka asahi shinbun*, November 22, 1927; and "Ryō mosu no kyozetsu de shōbōseki mo kōka suru," *Osaka mainichi shinbun*, November 26, 1927; and "Bōseki Rengōkai jiko yōgō no nayami," *Osaka asahi shinbun*, November 27, 1927, "Shinbun kiji," Vol. 14; and "Nōritsu zōshin wa kenchō," *Chūgai shōgyō shinpō*, October 9, 1928, "Shinbun kiji," Vol. 15; and "Tōyama, Ashikaga, Sankō mo Bōseki Rengōkai e kanyū," *Osaka mainichi shinbun*, April 24, 1930, "Shinbun kiji," Vol. 16.
37. "Bōseki sōtan wa enchō," *Osaka asahi shinbun*, January 28, 1928, "Shinbun kiji," Vol. 14.
38. "Honpō menshi kanzei mondai shiryō," *DaiNihon Bōseki Rengōkai geppō*, 1929 (444): 23–27.
39. "Menshokushi yunyū kanzei ni kansuru chinjō sho," *DaiNihon Bōseki Rengōkai geppō*, 1929 (445): 1–3.
40. "Menshi kyōchō no uchigeki wa shobōseki ni," *Osaka asahi shinbun*, November 5, 1929, "Shinbun kiji," Vol. 15.
41. "Sōgyō tanshuku jikkō ketsugi ni saishite," *DaiNihon bōseki rengōkai geppō*, 1930 (449): 2; "Bōseki wa ichiwarihō no sōtan wa hitsuyō da," *Osaka mainichi shinbun*, November 20, 1929, "Shinbun kiji," Vol. 15.
42. "Bōseki ketsugi sōtan wa futō," *Osaka jiji shinpō*, December 13, 1929, and "Bōseki no sōtan hantai," *Osaka mainichi shinbun*, January 14, 1930, "Shinbun kiji," Vol. 15.
43. "Bōseki sōtan ni hikōshiki keikoku," *Tokyo nichinichi shinbun*, January 29, 1930; "Bōseki gyōsha rōbai su," *Osaka mainichi shinbun*, January 30, 1930; "Bōseki sōtan ni rōdō jiken no jōgai rei," *Osaka asahi shinbun*, January 30, 1930, Vol. 15.
44. "Iinkai rokuji," *DaiNihon Bōseki Rengōkai geppō*, 1930 (449): 1–2.
45. "Bōseki no jikan enchō o yurusanu," *Osaka mainichi shinbun*, January 30, 1930, "Shinbun kiji," Vol. 15.
46. "Bōseki no sōgyō tanshuku wa zettai ni hantai suru," *Kōbe shinbun*, February 14, 1930, and "Tatsu mono issai ni bōseki no sōtan o kōgeki," *Osaka jiji shinpō*, February 4, 1930, "Shinbun kiji," Vol. 15.
47. "Honnen jōhanki kaku bōseki kaisha eigyō seiseki," *DaiNihon Bōseki Rengōkai geppō* 1930 (455): 33–37, and Kanebō Kabushiki Kaisha Shashi Hensanshitsu, *Kanebō hyakunen shi* (Osaka: Kanebō Kabushiki Kaisha, 1968), pp. 994–995. Even though Kanebō was one of the largest and most successful firms, its results were, for some reason, not included in the cited report in the August issue of the *Dai Nihon Bōseki Rengōkai geppō*.
48. "Tōkai sōkai ni okeru Abe iinchō ensetsu," *DaiNihon Bōseki Rengōkai geppō* 1930 (452): 2–3.
49. "Indo yunyū menpu kanzei zōritsu an teian ni saishi mengyō san dantai no seimeisho," *DaiNihon Bōseki Rengōkai geppō* 1930 (450): 3–5.
50. Shōji, *Bōseki sōgyō*, pp. 537–549.
51. Abe Fusajirō, "Indo menpu tokkei kanzei an no tsūka ni tsuite," *DaiNihon Bōseki Rengōkai geppō* 1930 (451): 31–32.
52. "Bōseki shinshijō kaitaku iyoiyo gutaika su," *Hōchi shinbun*, May 24, 1930, "Shinbun kiji," Vol. 16.
53. "Bōseki sōtan gofun kakuchō," *Osaka jiji shinpō*, May 22, 1930; "Sōtan no dai kakuchō," *Osaka mainichi shinbun*, May 25, 1930; and "Bōseki ichiwari sōtan kakuchō iyoiyo jisshi ni kesu," *Osaka asahi shinbun*, June 8, 1930, "Shinbun kiji," Vol. 16.

54. "Sōtan no dai kakuchō," *Osaka mainichi shinbun*, May 25, 1930, "Shinbun kiji," Vol. 16.
55. "Kishiwada bōseki sara ni tai sōtan dankō," *Osaka asahi shinbun*, June 25, 1930; "Bōsekikai wa taiseijō shizen sōtan ga okonawareyō," *Osaka jiji shinpō*, June 25, 1930; "Bōseki no jieiteki gensan tsui ni katabansei saiyō e," *Tokyo nichinichi shinbun*, June 26, 1930; "Dai bōseki kaisha ga katabansei saiyō o shunjun," *Osaka mainichi shinbun*, July 17, 1930, "Shinbun kiji," Vol. 16.
56. Nisshin Bōseki Kabushiki Kaisha, *Nisshin bōseki rokujū nen shi* (Tokyo: Keizai Ōraisha, 1969), p. 398; and Kanebō Kabushiki Kaisha Shashi Hensanshitsu, *Kanebō hyakunen shi*, pp. 221–225; and "Kishiwada bōseki sara ni tai sōtan dankō," *Osaka asahi shinbun*, June 25, 1930, "Shinbun kiji," Vol. 16.
57. "Bōseki sōtanritsu no kakuchō o mitomeru," *Osaka asahi shinbun*, August 21, 1930, "Shinbun kiji," Vol. 16.
58. "Sōtan kakuchō hantai ikensho teishutsu," *Osaka jiji shinpō*, September 9, 1930; "Bōseki sōtan kakuchō ketsugi," *Osaka jiji shinpō*, September 9, 1930, "Shinbun kiji," Vol. 16.
59. "Sakunen kahanki kaku bōseki kaisha eigyō seiseki," *DaiNihon Bōseki Rengōkai geppō* 1931 (462): 28–32, and Kanebō Kabushiki Kaisha Shashi Hensanshitsu, *Kanebō hyakunen shi*, pp. 994–995. Again, Kanebō's results were not included in the cited report in the *Dai Nihon Bōseki Rengōkai geppō*.
60. "Bōseki wa taiseijō shizen sōtan ga okonawareyō," *Osaka jiji shinpō*, June 25, 1930, "Shinbun kiji," Vol. 16.
61. "Mengyō no jūdanteki tōsei daiippō," *Tokyo nichinichi shinbun*, July 4, 1930, "Shinbun kiji," Vol. 16.
62. Nishikawa, *Nihon teikokushugi*, pp. 192–193, and Itō Chūhyōei, "Bōseki jigyō no tenkai wa shinkikai e," *Daiyamondo* October 21, 1931, *19*: 29–30, on p. 29.
63. Nisshin Bōseki Kabushiki Kaisha, *Nisshin bōseki*, p. 395, and Kanebō Kabushiki Kaisha Shashi Hensanshitsu, *Kanebō hyakunen shi*, p. 237.
64. Hashimoto, *Daikyōkōki*, p. 177.
65. "Bōseki kaisha kan ni mo sōtan kanwa ron okoru," *Tokyo nichinichi shinbun*, October 17, 1930, "Shinbun kiji," Vol. 16. "Kōritsu sōtan de bōseki ijō na bōri," *Tokyo asahi shinbun*, March 10, 1931; "Yunyū Shina shi no appaku de chūshōbō himei o agu," *Kōbe shinbun*, March 11, 1931; "Bōseki rengōkai kasanete sōtan kanwa o ketsugi sen," *Kokumin shinbun*, March 13, 1931; and "Fukeiki o yoso ni bōseki no rieki zōdai su," *Osaka mainichi shinbun*, March 15, 1931, "Shinbun kiji," Vol. 17.
66. Asahi Shinbun Keizaibu, *Asahi keizai nenshi, 1932* (Osaka: Asahi Shinbun Sha, 1932), pp. 146–147; "Bōseki rengōkai kasanete sōtan kanwa o ketsugi sen," *Kokumin shinbun*, March 13, 1931; "Shichigatsu irai no bōseki sōtan kanwa seishiki kettei," *Chūgai shōgyō shinpō*, April 19, 1931; "Mushi sareta menshishō no yōkyū," *Tokyo nichinichi shinbun*, April 21, 1931; "Bōseki, jūgatsu irai nijūwari sōtan ni omomuku," June 5, 1931, *Kokumin shinbun*; "Jūgatsu irai bōseki sōtanritsu ichiwari happun ni kettei," *Chūgai shōgyō shinpō*, July 7, 1931, "Shinbun kiji," Vol. 17.
67. "Honnen jōhanki kaku bōseki kaisha eigyō seiseki," *DaiNihon Bōseki Rengōkai geppō* 1931 (468): 32–37.
68. "Muan ni fueru bōseki no suisū," *Osaka asahi shinbun*, September 18, 1931, "Shinbun kiji," Vol. 17.
69. "Tōkai sōkai ni okeru Abe iincho ensetsu," *DaiNihon Bōseki Rengōkai geppō* 1931 (464): 2–3.

70. "Donata mo umisen yamasen," *Osaka asahi shinbun*, October 7, 1931, and "TaiEi mengyō sen—osoreru ni tarazu," *Osaka mainichi shinbun*, October 30, 1931, "Shinbun kiji," Vol. 17.
71. Tsuda Shingo, "Mengyō wa kinsaikinshi ga hitsuyō," *Daiyamondo* October 21, 1931, *19*: 28–29.
72. "Zaikai jōsei no seijika," *Daiyamondo* October 21, 1931, *19*: 7–8.
73. "Shōwa rokunen kahanki kaku bōseki kaisha eigyō seiseki," *DaiNihon Bōseki Rengōkai geppō* 1932 (474): 34–39.
74. "Kyōnenchū no honpō bōsekigyō," *DaiNihon Bōseki Rengōkai geppō* 1932 (478): 2–15, on pp. 11–12.
75. "Zenkoku menshi bōseki tōkei hyō," for September 1931 in *DaiNihon Bōseki Rengōkai geppō* 1931 (470): 1–38, on p. 1, and "Zenkoku menshi bōseki tōkei hyō," for December 1931, in *DaiNihon Bōseki Rengōkai geppō* 1932 (473): 1–38, on p. 1.
76. "Zenkoku menshi bōseki tōkei hyō," for December 1930 in *DaiNihon Bōseki Rengōkai geppō* 1931 (461): 1–36, on p. 1.
77. Abe Fusajirō, "Shōwa shichi nen no mengyōkai," *DaiNihon Bōseki Rengōkai geppō* 1931 (472): 17–18.
78. Seki Keizō, *The Cotton Industry of Japan* (Tokyo: Japan Society for the Promotion of Science, 1956), p. 311, and "Shōwa kyūnenchū no bōsekigyō," *DaiNihon Bōseki Rengōkai geppō* 1935 (512): 5–19, on p. 19. According to Bōren's statistics, only one firm had a loss in 1932 and 1933. See also "Shōwa sannen irai no saikō kiroku," *Osaka jiji shinpō*, January 30, 1933, "Shinbun kiji," Vol. 19.
79. Seki, *The Cotton Industry*, p. 311; "Bōseki sōtan ritsu," *Osaka mainichi shinbun*, June 9, 1934; and "Bōseki sōtan kanwa ni gyakkō no daizōsui," *Osaka mainichi shinbun*, June 26, 1934, "Shinbun kiji," Vol. 20. By this time, one holiday was estimated as equal to a production cut of 3.8%. See Asahi shinbun Keizaibu, *Keizai nenshi*, 1932, p. 146.
80. "Mannen sōtan o kaishō ichio sui'i o miyo," *Osaka asahi shinbun*, July 6, 1934, "Shinbun kiji," Vol. 20.
81. "Bōseki no sōtan gofun kakuchō," *Osaka mainichi shinbun*, February 2, 1935; "Shichigatsu irai no sōtan kakuchō kettei," *Osaka mainichi shinbun*, May 12, 1935; "Sennai de zōsui keikaku," *Kyōsei nippō*, May 19, 1935, in "Shinbun kiji," Vol. 20.
82. "Sōtan hōhō no kaikaku ga mushiro senketsu mondai," *Tokyo nichinichi shinbun*, July 20, 1935, in "Shinbun kiji," Vol. 20.
83. "Riron no binkonka," *Tokyo nichinichi shinbun*, July 30, 1935, in "Shinbun kiji," Vol. 20.
84. "Iinkai rokuji," *DaiNihon Bōseki Rengōkai geppō* 1935 (517): 1–3, on p. 2, and "Angai assari zōsui yokuseian naru," *Osaka asahi shinbun*, July 26, 1935, in "Shinbun kiji," Vol. 20.
85. "Iinkai rokuji," *DaiNihon Bōseki Rengōkai geppō* 1936 (526): 1–4.
86. "Kyōnenchū no honpō bōsekigyō," pp. 12–13.
87. Hirasawa Teruo, "Shōwa kyōkōka ni okeru jūyō sangyō tōsei hō no unyō ni kansuru ichi kōsatsu," *Rekishigaku kenkyū* 1991 (619): 1–18, p. 14.
88. The issue of whether or not to apply for legal status remained controversial in 1935. See "Bōseki rengōkai no hōteki karuteruka," *Kōbe shinbun*, July 28, 1935, "Shinbun kiji," Vol. 20.
89. Hirasawa, "Shōwa kyōkōka," pp. 14 and 16.

90. Miyajima Hideaki, "1930 nendai Nihon no dokusen soshiki to seifū," *Tochi seido shigaku* 1986, *28*: 1–23, on pp. 8–9.
91. "Bōseki no sōtan ritsu gofun hachi ri o kakuchō," *Osaka mainichi shinbun*, October 17, 1931, "Shinbun kiji," Vol. 17.
92. "Sōtan ritsu o kimeru," *Osaka asahi shinbun*, October 30, 1932, and "Bōseki to menshishō hanmoku sen'eika," *Osaka mainichi shinbun*, November 29, 1932, "Shinbun kiji," Vol. 19.
93. Y.T. Sei, "Mengyō tōsei no genjō," *DaiNihon Bōseki Rengōkai geppō* 1935 (516): 4–8, on pp. 4–6. The Important Export Industries Association Law of 1925 and the Industrial Association Law of 1931 gave industrial associations legal status. Some members of the Spinners Association that produced cloth could also belong to an industrial association and thus to Menkōren.
94. Y.T. Sei, "Mengyō tōsei," p. 6. The Export Association Laws of 1925 and 1931 gave legal status to the export associations.
95. Y.T. Sei, "Mengyō tōsei," pp. 7–8; "Menpu yushutsu tōsei ni Bōren danyo hantai su," *Osaka mainichi shimbun*, March 24, 1935; "Bōren kameisha yushutsu kumiai ni kanyū," *Osaka mainichi shinbun*, June 13, 1935; and "Jōsei no henka kara Bōren to menkōren tsui ni teikyō e," *Osaka asahi shinbun*, July 28, 1935, "Shinbun kiji," Vol. 20.
96. For example, see "Mengyō Chūō Kyōgikai rokuji," *DaiNihon Bōseki Rengōkai geppō* 1935 (519): 1.
97. "Jūyō taisaku chōsa iinkai rokuji," *DaiNihon Bōseki Rengōkai geppō* 1936 (523): 1–3, on pp. 1–2, and "Dai yonjū kyūkai teiki sōkai," *DaiNihon Bōseki Rengōkai geppō* 1936 (524): 1–4, on p. 1.
98. "Yushutsu tōsei kitei chū nyūsatsu seido kaisei ni kansuru chinjōsho naranni ikensho," *DaiNihon Bōseki Rengōkai geppō* 1935 (517): 3–5.
99. "Bōeki tōsei ni kansuru Nihon Keizai Renmeikai ikensho," *DaiNihon Bōseki Rengōkai geppō* 1936 (530): 9–10, on p. 10.
100. In May 1946 the Nihon Bōseki Dōgyōkai was formed, and in April 1948 it became the the Nihon Bōseki Kyōkai.

Index

Abe, Fusajirō (head of the Japan Spinners Association), 212, 214, 215, 222, 223
  response to the Great Depression, 217, 218, 219, 220, 221
abortion, 73, 79
acquired characteristics, 64
Adachi, Yoshiyuki, 189
Adams, Mark, 77
adoption, 62, 65
Ainu, 4, 85–93
air raids, 201
Akasaka Filature, 144, 145, 146, 154
alcohol, 69, 73, 74
alcoholism, 66, 72, 73, 74
America Japan Sheet Glass Company, 6, 175–6
ancestor worship, 70
Anderson, William Edwin (1842–1900), 22
anthropology, 84, 85–7, 97
architectural history, 195
architecture
  traditional Japanese, 161–2, 164
  Westernization, 162–3, 164
  windows in, 163–4, 177–8
Army, 15, 31–2
Asahi Glass, 174, 175, 176
Asakura, Seiichi, 188
Austria, 41

bacillus, 13, 27–8
bacteriology, 13, 25, 31
Baelz, Erwin von (1849–1913), 20, 23–4, 64, 85–6
bakufu, 136, 152
Banichi, Yasuhirō, 104
Bartholomew, James, x, 2, 76, 77, 78
Bauduin, Antonius (1822–85), 19
Beard, George M., 40–1
Beijing Tariff Conference, 209, 210
beriberi, 3, 13–32
  hospital, 17–19, 23
Berlin, 65, 66, 76, 78
  University, *see under* universities
Big Five (Construction Companies), 186
biological, 61, 63, 64, 65, 70; *see also* biology *under* sciences
biological determinism, 70, 75; *see also* Mendelian laws
biologists, 65, 69, 76
Birch-Hirschfeld, Felix Victor, 24
bodies, 61–77
  female, 61, 64, 67, 68, 69, 74, 75, 76, 77
  human, 63, 66, 72
  Japanese, 62, 63, 64, 70
  male, 39, 66
  state-body analogy, 72–3
bodily improvement (*taishitsu kairyō*), 64, 66–8, 70, 71, 76; *see also* improvement *under* constitution
body politic, 39
borrowed wombs, 73
breeding, selective, 64; *see also* crossbreeding *under* races; heredity
bricks, 162
Britain, Great, 2, 45
Brunat, Paul, 6, 137, 138, 139, 140, 142, 145, 150, 151, 153, 154
Brunton, Richard Henry, 165
Buddhism, 70
bunmei kaika (civilization and enlightenment), 2, 139, 146, 152
Bureau of Hygiene, 14–15, 26–7
bushidō (the way of warriors), 70, 74, 80

cement, 162
Chaiklin, Martha, 6, 7
chambon, 146, 148, 149
Chance Brothers Glass and Chemical Works, 165, 170

Charcot, Jean-Martin, 41
China, xiii, 8, 44–5
  Japanese mills in (*zaikabō*), 212, 215
  as a market for Japanese textiles, 209, 210, 216, 220, 221, 222
  as rival for Japanese spinning companies, 209, 212, 213, 215, 220, 224
  War (1937–45), 226–7
Chinese, 4
"choice of technique," 141, 150–1, 155
cholera, 14, 116–17
Christian, 74, 75
civilization/civilized, 61, 66, 67, 79
Clancey, Gregory, 6
class, 61, 62, 64, 67, 68–9, 70, 73, 80
  middle, 64, 67, 68–9, 73, 80
  upper, 67, 68–9, 70
  warrior, 62
clothing, 63, 64
colleges, 65, 71–2, 74, 75
  department of race improvement, 71
  faculty of medicine, 71
  Japan Women's College, 65, 71–2
  women's, 65
Commerce and Industry, Ministry of, 224–6
Confucianism, 70
constitution, 63, 66, 67, 70–1, 75
  improvement, 66, 70–1, 75
  see also bodily improvement; eugenics; improvement *under* race
contractor, 186
cotton industry, 7
Council on Tariffs (in Japan), 215, 217
crime, 66
criminal anthropology, 66

daiku, 6, 7, 183–202
Dai-Nippon Spinning Company, 223
Dairen, xiv, 103, 106, 112, 123
  city plans, 107–8, 109, 111–12
  waterworks, 118–19, 120
Dajōkan, 194, 201
danson johi, 73
Decoration Bureau, 22
degeneration, 66, 68, 73, 75, 79, 80
detective investigation, 67–8, 79
diet, 31, 63–4, 75
Diet, the, 209, 216, 225
Disse, Joseph (1852–1912), 28
Drei Abhandlung zur Sexualtheorie, 45
Dresser, Christopher, 162
drinking, 63, 69, 74; *see also* alcohol; alcoholism; temperance
Du Bois-Reymond, Emil, 62
Du Bousquet, Albert Charles, 137
dysentery, 115

earthquake, Kantō, 121
Economy, Trade and Industry, Ministry of, (M.E.T.I.), xiii
Edo, 2, 62
  castle, 187
  see also Tokyo
education, 63, 65, 67, 69, 71–4, 76–7
  Ministry of, 197–9
  physical, 63
  women's, 65, 69, 71–4
  see also colleges; universities
educators, 61, 63, 65, 69
electrical hand-tools, 201
Ellis, Havelock, 38, 50
emperors, 80; *see also* imperial
empire, 42, 50
England
  abandonment of gold standard by, 221–2, 226
  spinning industry as negative model, 223
  as trade rival, 210, 217–18
environment, 64, 68–70
eugenics, 61–3, 65–6, 68, 70–6, 77–80, 84
  associations, 70
  hybrid, 76
  institutionalization of, 66
  laws, 70, 72, 74–6
  negative, 74
  professionalization of, 65
  scientification of, 68
  transplanting, 71
  see also betterment *under* race; bodily improvement; improvement *under* race; National Eugenics Law
Eugenics Education Society (London), 66
evolution, 63, 76, 78
  theory of, 63, 76

exhibitions, international, 140, 144, 154
export associations (in Japan), 224–5

factories, 104, 106
Factory Law (1911 & 1923), 209, 217
families, 62, 67, 68, 70, 73, 74, 76
  patriarchal, 68, 73
  state, 70
  system, 68, 70, 74, 76
feminists, 61, 72–6, 80
fitness, 61, 65, 66–7, 69, 70–3
  biological, 70
  intercourse, 66–7
  mental, 65, 69
  moral, 65, 69
  physical, 65, 69, 72
  reproductive, 66–7
Fletcher, W. Miles, 7
Fordism, 184
Foreign Affairs, Ministry of, 218
Forel, August, 45
forest, 185, 186
France, 2, 41, 45
free love, 70; *see also* love *under* marriage
Freud, Sigmund, 41, 45
Frühstück, Sabine, 3, 4, 5, 77
Fujikawa, Yū, 45–6
Fujino Yutaka, 78–81
Fujitani, Takashi, 2
Fujiyama Tanehiro, 169
Fujo shinbun, 48
fukoku kyōhei, 1, 14, 135
Fukushima Spinning Company, 214, 217
Fukuzawa Yukichi, 61, 63, 78
Fuller, George A., and Company, 121
Futsū mokkō jutsu, 198, 199

Galton, Francis, 61, 63, 65, 67, 78
"Garden City" movement, 108, 112, 117, 119
Garon, Sheldon, 81
Geisenheimer, F., 136–8
gender, 61, 65, 68, 72, 77, 80
  ideology, 65, 80
genetics, *see under* medicine
Geoltz, Friedlich Leopold, 62
Germany, 26, 40–1, 45, 64–6, 67, 76–9; *see also* Berlin; Freiburg *and* Strassburg *under* universities
germ plasm, 64–5, 69
germ theory, 25–6, 28, 30–1
gijutsu, 183
gijutsu rikkoku, 5
Ginza, 164, 194
glass, 6, 8
  demand, 170, 173, 178
  Edo period, 163
  exports, 175
  government concern, 167–8, 171
  imports, 163–4, 165–7, 169
  production techniques, 163, 174–5
Gluck, Carol, 2
gold standard
  as an issue in Japan, 216, 221
gonorrhea, 66, 72
good wife, wise mother, 65; *see also* ideology *under* gender
Gordon, Andrew, 81
Gosankai, 211–12, 216, 218–19
government, 62, 65, 74, 75
  Meiji, 62
  see also state
Great Depression, 7
  impact on Japan, 207–9
  impact on Japanese spinning industry, 221–2, 225–6
Greater East Asia Co-Prosperity Sphere, 8
Greater Japan Private Women's Hygienic Association, 71, 78
Great-Japan Private Society for Hygiene, 21, 28
Guandong (Kwantung)
  Army, 107, 124
  Leased Territory, 106
guild, 185
gynecology, 37

Habuto, Eiji, 37, 39, 42, 50
Hakodate, 186
Hamaguchi, Osachi, 215
han, 136, 152, 154
Handbuch der Neurasthenie, 40
Hanley, Susan B., 161
Hara Akira, 208
Harada, Yutaka (?–1894), 22
Hashimoto, Tsunatsune (1845–1909), 16, 22
Hattori Shōten, 213

Hayami Kenzō, 142, 144, 146, 151–4
Hayashi, Tadahiro, 188
Hayden, van der, 22, 25
Hirasawa, Teruo, 224
Hiroi, Komaji, 26
health certificates, 61, 67, 72
  prenuptial, 61, 67
Hearn, Lafcadio, 162
Hècht, Lilienthal and Company, 137, 138, 140
Helmholtz, Hermann von, 62
heredity, 63, 64, 65, 69–70, 75, 79
  environmental approaches, 64, 69
  see also acquired characteristics; genetics *under* medicine; germ plasm, inheritance; Mendelian laws; natural selection
Hirano, Kōsuke, 171–2
Hiratsuka, Raichō, 72–4, 76, 80; *see also* feminists
Hiroshima, 198
Hirschfeld, Magnus, 41
history of technology, 183
Hōchi newspaper, 67, 68, 73
Hokkaido, 189
Holland, 1, 2
home, 64, 72, 76
  economics, 72
Home Affairs, Ministry of, 217–18
homosexuality, 37, 49
Hoppe-Seyler, Felix, 62
Horiuchi, Toshikuni (1844–95), 22
House of Peers, 63, 74, 75
House of Representatives, 73–5
housing, 63
hysteria, 41, 43, 51

ideas, 61–3, 73
  adaptation of, 61
  transplantation of, 62
  Western, 63
identity, 64, 80
Ikeda, Kensai (1841–1918), 17–18, 22
Imamura, Ryōan, 17
Imbry-kan, 190
immigration, 112–13
imperial, 63, 70–1, 74
  decree, 63
  institutions, 71, 74
  palace, 196
  university, 195–8
  see also emperors
Important Industries Control Law (IICL), 208
  impact on Japan Spinners Association, 223–4, 226
India
  as a market for Japanese textiles, 220
  as supplier of raw cotton, 214
  trade tensions with, 209, 210, 215, 217–18, 225–6
Indonesia, 30
influenza, 116
inheritance, 63–4, 69
  of talents, 63
Inoue, Junnosuke (Minister of Finance), 207–8, 216, 221
Inoue, Kaoru, 139, 153, 166
Inoue, Tetsujirō, 63, 64
Interior, Ministry of the, 14–15, 26
Ise, Jōgorō (1852–?), 22
Ishiguro, Tadanori (1845–1941), 16, 22–3, 28
Itō, Chuta, 195
Itō, Hirobumi, 136–7, 139
Itō, Keishin, 170–1, 173
Itō Takenosuke, 223
Iwaki Glass, 171
Iwaki Tasujirō, 171
Iwakura Mission, 135, 137, 165
Iwasaki, Toshiya, 173–4

Janes, L.L., 162
Japan Economic Federation (Nihon Keizai Renmeikai), 225
Japan Federation of Cotton Woven Goods Industrial Associations (Menkōren), 224, 225
Japan Federation of Cotton Yarn Merchant Associations, 224, 225
Japan Knitted Goods Export Association, 218, 219
Japan Spinners Association (DaiNihon Bōseki Rengōkai, abbreviated Bōren), 7
  attitude toward the Important Industries Control Law, 223–4, 226
  overall role, 209, 211, 216, 225–7
  postwar emergence, 227, 232n100

and production curtailments, 211–16, 216–24, 226
rivalry with other business groups, 224–5, 232n93
and tariff reform, 215, 217
and trade disputes with India, 210, 217–18, 226
Jarves, James Jackson, 161–2
Jidō kenkyū, 45
jinshu, 72, 74
Johnston, William, xiv

Kaitakushi (Colonization Ministry), 189
Kajima, 186, 187
Kamio, Akira, Lt-General, 106
Kanegafuchi Spinning Company (Kanebō), 214, 217, 219, 221, 229n47
Katō, Hiroyuki (1836–1916), 27, 63–4, 78, 80
Kishiwada Spinning Company, 219
Kekkon shinsetsu, see Tsūzoku kekkon shinsetsu
Kenchiku-ka, 195
Kenchiku zasshi (*Architecture Journal*), 194
Kenkō, 161
Kevles, Daniel, 79
Key, Ellen, 72
Kigo, Kiyoyoshi, 195
Kiku-jutsu, 192–3, 199, 201
Kitasato, Shibasaburō (1852–1931), 14, 29
knowledge
scientific, 39
Kobayashi, Tan (1847–94), 17–18
Kobe, 186–7
Kōbudaigakkō, 194, 197
Kobushō, 188
Kōbushō kankōryō, see Akasaka Filature
Koch, Robert, 24–6, 28–9
"*Kōfuku naru kekkonhō*," 67
Koganei, Yoshikiyo (1859–1944), 86–7
Kōgyō Gakkō, 200
Kōgyō Iken, 135
Korean War, 201
Koura, 189
Krafft-Ebing, Richard von, 37, 50
Kuni no hikari, 73, 81
Kuratsuka, Yoshi, 108
Kwantung, *see under* Guandong (Kwantung)

League for the Protection of Mothers and Sexual Reform (Germany), 67
Le Bon, Gustave, 68
leprosy, 66–7
Libbey-Owens Sheet Glass Company, 6, 175–6
Lock, Margaret, 76–7
Löffler, Friedrich (1852–1915), 26
Low, Morris, 76, 77, 78, 80

Maebashi filature, 142–3, 146, 152
Maeda, Masana, 135
malaria, 23
Manchuria, xiv, 5
Manchurian Incident (1931), 8, 220, 226
Manderson, Lenore, 114
Mantetsu, *see* South Manchuria Railway Company (SMR)
marriage, 61–5, 67–8, 70–4, 76, 78, 79
arranged, 62, 70, 73, 74, 76
de facto, 73
eugenics laws, 72
individualistic, 72
inter-, 63
love, 68, 72; *see also* free love
mixed, 64, 65, 70, 78
racial, 72
social, 72, 73
see also crossbreeding *under* races; health certificates
masculinity, 40–2, 44
masturbation, 37, 41–3, 47–9, 54
Matheson, R.O., 123
Matsubara, Yōko, 76, 79
Medical Affairs, Office for, 14
medical authorities, 71, 74; *see also* scientific authorities
Medical Newspaper, 16, 19
medicalization, 61, 67–8, 70, 75–7
of life, 61, 77
of marriage, 61, 67–8, 76
of race improvement theory, 70, 75
medicine, 62–6, 69–80
bacteriology, 79
basic, 76

medicine—*continued*
Chinese-style, 6, 15–17, 19
dermatology, 79
Dutch, 19
European, 62
genetics, 69, 71–2, 76, 79
hospital, 13, 17, 29
hygiene, 62
internal, 64
Kanpō, 14–18
laboratory, 13
medical chemistry, 62
modern, 66
obstetrics, 79
pathology, 64, 66
pharmacology, 62
physiology, 62, 65, 70, 74, 75, 76, 77, 78, 80
psychiatry, 66, 79
scientific, 13, 30
sexology, 63, 77
social hygiene, 66, 79
Meerdevort, Johannes L.C. Pompe van (1892–1908), 18–19
Meiji, 62, 63, 68, 71, 75–6, 79–80
Emperor, 2, 15, 21
Empress, 15
Gakuin, 190
Government, 141, 150–1, 153
reforms, 63
Restoration (1867–8), 1, 14–15, 32, 62, 135, 186, 192
Mendelian laws, 65, 69; *see also* biological determinism
menstruation, 83, 87–96
Meyer, Julia, 65
Miasma, 17, 22–3, 28, 31
military, 40, 47
administration, 43
decline, 45
medicine, 44
physical, 42
uniform, 40
Minbushō, 137, 139, 153, 155
Minseitō, 216, 221
minzoku, 70, 72, 79, 80, 81; *see also* Yamato *minzoku*
Mito, 168
Mitsubishi "Zero" Fighter, 8
Mitsui Bussan, 166
Miya-daiku, 185, 187, 196
Miyajima, Hideaki, 224
Miyajima, Seijirō, 212
Miyake Hiizu (1848–1938), 17–18, 22, 24
modernization, 13–14, 31–2, 61, 74, 81
Monist League (Germany), 67
Mori, Arinori, 165
Mori, Ōgai, 47
Morita, Shoma, 44
Morris-Suzuki, Tessa, xiii, xiv, 5, 77, 81
Morse, Edward S., (1838–1925), 84
Mueller, Casper, 142, 151, 153
Mukō, Gunji, 49
Müller, Franz Carl, 40
munitions, 8

Nagai, Hisomu (1876–1957), 70, 77, 80
Nagasaki, 1, 62, 186
Nagasaki kaisho, 166
Nagoya, 187
Nagayo, Sensai (1838–1902), 14, 26–7
Naimushō (Ministry of Home Affairs), 144
Nakamura, Takafusa, 208
Nakatani, Norihito, 192
Naruse, Jinzō, 65, 69, 71–2, 75–6
National Eugenics Law (1941), 5, 70, 75
nationalism, 71
nationalistic, 63, 67, 70
nation-building, 61, 65, 73
Natsume Sōseki, 178
natural selection, 66
navy, 15, 21, 23
NEC, xiii
Nemoto, Shō, 73–4, 75–6, 81
nervous system, 68–9
female, 68
neurasthenia, 3, 37, 39–41, 43–4, 51–4
Nichi-bei Ita Garasu Kabushiki Gaisha, *see* America-Japan Sheet Glass Company
Nihon Garasu Gaisha, 170–1
Niigata, 190
Nikkō, 187
Nishikawa, Hiroshi, 212
Nishimura Katsuzō, 169–70
Nishio Suehiro, 217

Nisshin Spinning Company, 212, 219
Nōshōmushō (Ministry of Agriculture and Commerce), 135, 154
nutrition, 14, 21–2, 44, 51, 64, 69, 70
 Mal-, 69
 see also diet

Ōbayashi, Ukonji, 62; *see also* Ōsawa, Kenji (1852–1927)
Oberländer, Christian, 3
Odaka, Atsutada, 139, 145, 15–4
Ōgata, Masanori (1855–1919), 13, 22–3, 26–30
Ōkuma, Shigenobu, 2, 5, 9, 47, 137–8
Okura, 186
Ōkurashō (Ministry of Finance), 137, 152–3
Osaka, 2, 14, 171, 186, 190
Osaka Gōdō Spinning Company, 212
Ōsawa, Gakutarō (1863–1920), 65
Ōsawa, Genryū, 62
Ōsawa, Kenji (1852–1927), 4, 22, 61–81
 collaboration with Hiratsuka Raichō, 72–4, 76
 collaboration with Naruse Jinzō, 65, 71–2, 74, 76
 collaboration with Nemoto Shō, 73–4, 76
 early life of, 62–3
 genetic ideas of, 69–70
 on marriage, 64–71
 and medicalization of race improvement theory, 61, 67–71
 as physiologist, 61, 69, 70, 74–5, 77
 as scientific authority, 75
 as statist, 73–4
 as Tokyo University Professor, 62–3, 70, 73
Oshima, Mitumoto, 189
Oslers, 165
Otsubo, Sumiko, 4, 5
oyatoi gaikokujin, 62, 190

Pacific War (1941–5), 8
Pasteur, Louis, 25
Pauer, Erich, xiii
Perrins, Robert, 5
Perry, Commodore Matthew C., 1, 2
Pettenkofer, Max von (1818–1901), 19, 26
physicians, 61, 72
physiologists, 61–2, 69–70, 74–5, 77; *see also* physiology *under* medicine
Pioneer, xiii
pneumonic plague, 109–10
poison, 16, 18–19
Port Arthur (Ryojun), 106, 116
potency, 39
production curtailments (*sōgyō tanshuku*, abbreviated *sōtan*), 211
 1927–1929, 211–16
 after, 1930, 216–24
 overall assessment, 226
program of industrialization, *see* *shokusan kōgyō*
psychiatry, 37, 39, 48
psychology, 37
pedagogy, 37, 39, 48
public health, 106–7
Public Works, Ministry of, 7, 197, 198, 199

quarantine, 74, 113
quinine, 16

race(s), 42, 44–5, 61, 63–4, 66, 68–76, 77–9, 83–90, 92–4, 96–7
 betterment, 63, 64
 Caucasoid, 63
 commercialization of, 68
 crossbreeding, 63–4
 human, 66
 improvement, 61, 63, 64, 66, 68, 69, 70–3, 75–6
 inferior status of, 61, 63, 64, 70, 76
 Japanese, 61, 63, 64, 70, 74, 76, 78
 marriage, 72
 mongoloid, 63
 popularization of, 68
 pure, 64; *see also minzoku*
 white, 63, 71; *see also* whitening
 yellow, 71; *see also Yamato minzoku*
racial, 61, 66, 69, 71–2, 80
 degeneration, *see under* degeneration
 difference, 80
 hierarchy, 61

racial—*continued*
marriage, 72
stock, 69
supremacy, 71
Racial Hygiene Society (Berlin), 66
Ragsdale, Kathryn, 67, 79
reformers, 61, 73, 74, 75
Christian, 74, 75
social, 61, 73, 74, 75
reforms, 63, 64, 67, 71, 75
reproduction, 61, 63, 64, 66–9, 70, 72, 75
revolution, 54
rice, 22–3, 25, 31
Rich Nation, Strong Army, *see fukoku kyōhei*
Russia, 2
Ryūjō, 21
Ryūkyū, 83, 87–90, 92–3
Ryūkyūan, 4

Sahkarov, Vladmir, 107–8, 109
Sakujikata, 188, 193, 196
Samuels, Richard J., 5
Samurai, 62, 70, 78, 80; *see also bushidō* (the way of warriors); warrior *under* class
sanitation bylaws, 114, 115
Sano, Tsunetami, 145, 154
Sapporo, 189–90
Sapporo Hokaikan, 189
Sasaki, Tōyō (1838–1918), 17
Satomi, Giichirō, 19
Sawada, Junjirō, 37
science of sex, 39
sciences, 61–2, 64–6, 69, 72, 75, 76–8, 80
biology, 64–5, 69, 76
cytology, 64
physics, 62, 77
Western, 61
Zoology, 65, 76, 78
see also eugenics
scientific authorities, 61, 74–5; *see also* medical authorities
Screech, Timon, 1
Scriba, Julius (1848–1905), 22
security of the nation, 37
seigaku, 39
seikagaku, 39
Sei no shinri, 37–8
Seirgakuijō yori mitaru fujin no honbun, 67
Seiyoku to jinsei, 37
Seiyūkai, 221
Sekkei/Sekō, 187
sex, 66, 67, 68, 70, 72, 77, 79
sexology, 37
see also under medicine; science of sex
sexual
abstinence, 39
behavior, 39, 50, 54
desire, 44, 51–2
dysfunction, 50
health, 51
immorality, 39
knowledge, 47, 50
organs, 51–2
perversion, 41
practices, 48
sexuality, 73
Shakaiteki eisei taishitsu kairyōron, 66, 67, 70–2, 79, 80–1
Shibata, Tsuguyoshi (1850–1910), 26
Shibusawa, Eiichi, 137–9, 141, 152, 154
Shimada, Magoichi, 170, 173–4
Shimada Glassworks, 173
Shima Trading Company, 175
Shimizu, 186–8
Shimizu Keiichi, 197
Shinagawa Glassworks, 168–71
Shinto, 62, 70, 186, 188, 195
Shogunal Institute of European Medicine, 62
Shokkō Gakkō, 197, 200
shokusan kōgyō, 1, 135–6, 140
Shufu no tomo, 54
silk industry, 6
smallpox, 3
social Darwinism, 5, 64, 71, 83–5, 88–9, 94–5, 97
soldier, 42
Sotomi, Tetsujirō, 218–19
South Manchuria Railway Company (SMR), 107, 114–15
hospital, Dairen, 103–4, 120–3
railworks Shahekou, 104, 106
soya bean trade, 110, 113
Speed, James, 169, 173
Spencer, Herbert, 84–5, 97

state, 62, 70, 72–5, 79–81
  family, 70
  Japanese, 74
  see also government
State Council, *see Dajōkan*
steam
  engines, 1, 139, 145
  reeling, 139, 144
Stepan, Nancy, 77–8
sterilization, 72, 74
Study in the Psychology of Sex, 38
subcontractors, 188
Sugita, Yosaburō, 175–6
Sugiura, Yuzuru, 139–41
suicide, 42
Suzuki, Zenji, 76–8, 80–1
syphilis, 66, 72

Tachikawa, Tomokata, 188–9
Taisei, 186
Taishitsu kairyōron, see Shakaiteki eisei taishitsu kairyōron
Taishō, 71
Taiwan, 83, 87–93
Taiyō, 48
Takagi, Kanehiro (1849–1920), 20–3, 28–9
Takahashi, Korekiyo (Minister of Finance), 208–9, 221
Takahashi, Yoshio, 63–4, 78
Takayama Kentarō, 171
Takenaka, Komuten, 186–8
Takenaka, Touemon, 187
Taniguchi, Fusazō, 212
Tanizaki Junichirō, 164, 177
Tariffs, 165–6
Tatsuno, Kingo, 196
tavelle, 146, 147, 149
Taylor, Wallace (1835–1923), 25, 27
Taylorism, 184
technology transfer, 6, 135–6, 190
telegraph, 1
temperance, 61, 73–5
Terada Tatsumi, 210
Terazawa, Yuki, 4–5, 76, 80
Tiegel, Ernst, 62
Tōda, Chōan (1819–89), 17
Todani, Ginzaburō, 104
Tōkaidō, 189
Tokugawa, 62, 67, 80–1, 139, 153, 155
Tokugawa, Iemochi (1846–66), 14
Tokugawa, Iesada (1824–58), 14
Tokyo, 2, 62
  Academy, 64
  rebuilding, 201
  University, *see under* universities
  see also Edo
Tokyo Medical Journal, 16, 24, 28
Tomioka Silk Filature, 5, 135, 138–9, 141–2, 144–6, 150–5
Torii, Ryūzō (1870–1953), 86–7
Tōyō gakugei zasshi, 64
Tōyō Glass Manufacturing Company, 176
toyomori see chambon
Tōyō Spinning Company (Tōyōbō), 212
traditions, 62, 74, 76, 81
  invented, 76, 81
Traweek, Sharon, 76, 77
treaties, 162
Tsuboi, Shōgorō (1863–1913), 86–7
Tsuda, Mamichi, 167
Tsuda, Shingō, 221
Tsukiji, 1, 187
Tsukuba, 21
Tsūzoku kekkon shinsetsu, 67–73, 78–81
tuberculosis, 25, 66–8, 115–16

Uesugi, Senpachi, 175
Ukeoi-shi, 186–8, 190, 193
undesirability, 63, 72
unequal treaties, 63–4
unfitness, 61, 66, 69, 72, 74; *see also* undesirability
United States of America, xiii, 41, 45
  Stock market crash in, 216
  as supplier of raw cotton, 210, 214
universities, 62–5, 70–1, 73–5, 77
  Berlin, 62
  faculty of medicine, 63
  Freiburg, 64, 65, 79
  Leipzig, 26
  Strassburg, 62
  Tokyo, 62–5, 70, 73, 74, 77
  see also colleges
U.S.–Japan Treaty of Amity and Commerce, 2
Utagawa, Yoshitora, 1

venereal diseases, 66, 68, 72, 73; *see also* gonorrhea; syphilis
Versuch über die Nervenkrankheiten, 40

Walton, Thomas, 168, 169
Wangjiatian reservoir, 118
war, 65, 67, 71, 74
  Russo-Japanese (1904–5), 31, 67, 71, 74, 106, 107
  Seinan Civil (1877), 15
  Sino-Japanese (1894–5), 31, 65, 67
  World War I, 43, 67
  World War II, 52, 201
wasan, 192
Watanabe, Kanae (1858–?), 25
wayō setchū, 7, 189–91, 199, 200
Weindling, Paul, 78–9
Weismann, August, 64–5, 69
Wernich, Agathon, 18
Western diet, 42
whitening, 63, 78; *see also* white *under* races
Witte, Sergei, 107
Wittner, David, 5–7
women, 61, 64–70, 72–6, 78, 80; *see also* education; female *under* bodies

Yamamoto, Senji, 50
Yamao, Yōzō, 138
Yamato *minzoku* (Japanese race), 69–70
Yamazaki, Masashige (1872–1950), 4, 83, 87–96
Yamazaki, Motomichi, 28
Yasuda, Tokutarō, 50
yellow scare, 71
Yokohama, 2, 186, 189–90, 201
Yokosuka Navy Yard and Iron Works, 188–9

zaguri, 138
Zaibatsu, 186, 189, 191, 193, 202
Zairai-kōhō, 184
Zōkagaku-shi, 195